TRANSIT COOPERATIVE RESEARCH PROGRAM

TCRP REPORT 113

Using Archived AVL-APC Data to Improve Transit Performance and Management

Peter G. Furth
NORTHEASTERN UNIVERSITY
Boston, MA

WITH

Brendon Hemily
HEMILY AND ASSOCIATES
Toronto, Canada

Theo H. J. Muller
DELFT UNIVERSITY OF TECHNOLOGY
Delft, The Netherlands

James G. Strathman
PORTLAND STATE UNIVERSITY
Portland, OR

Subject Areas
Public Transit

Research sponsored by the Federal Transit Administration in cooperation with the Transit Development Corporation

TRANSPORTATION RESEARCH BOARD

WASHINGTON, D.C.
2006
www.TRB.org

TRANSIT COOPERATIVE RESEARCH PROGRAM

The nation's growth and the need to meet mobility, environmental, and energy objectives place demands on public transit systems. Current systems, some of which are old and in need of upgrading, must expand service area, increase service frequency, and improve efficiency to serve these demands. Research is necessary to solve operating problems, to adapt appropriate new technologies from other industries, and to introduce innovations into the transit industry. The Transit Cooperative Research Program (TCRP) serves as one of the principal means by which the transit industry can develop innovative near-term solutions to meet demands placed on it.

The need for TCRP was originally identified in *TRB Special Report 213—Research for Public Transit: New Directions,* published in 1987 and based on a study sponsored by the Urban Mass Transportation Administration—now the Federal Transit Administration (FTA). A report by the American Public Transportation Association (APTA), Transportation 2000, also recognized the need for local, problem-solving research. TCRP, modeled after the longstanding and successful National Cooperative Highway Research Program, undertakes research and other technical activities in response to the needs of transit service providers. The scope of TCRP includes a variety of transit research fields including planning, service configuration, equipment, facilities, operations, human resources, maintenance, policy, and administrative practices.

TCRP was established under FTA sponsorship in July 1992. Proposed by the U.S. Department of Transportation, TCRP was authorized as part of the Intermodal Surface Transportation Efficiency Act of 1991 (ISTEA). On May 13, 1992, a memorandum agreement outlining TCRP operating procedures was executed by the three cooperating organizations: FTA, the National Academies, acting through the Transportation Research Board (TRB); and the Transit Development Corporation, Inc. (TDC), a nonprofit educational and research organization established by APTA. TDC is responsible for forming the independent governing board, designated as the TCRP Oversight and Project Selection (TOPS) Committee.

Research problem statements for TCRP are solicited periodically but may be submitted to TRB by anyone at any time. It is the responsibility of the TOPS Committee to formulate the research program by identifying the highest priority projects. As part of the evaluation, the TOPS Committee defines funding levels and expected products.

Once selected, each project is assigned to an expert panel, appointed by the Transportation Research Board. The panels prepare project statements (requests for proposals), select contractors, and provide technical guidance and counsel throughout the life of the project. The process for developing research problem statements and selecting research agencies has been used by TRB in managing cooperative research programs since 1962. As in other TRB activities, TCRP project panels serve voluntarily without compensation.

Because research cannot have the desired impact if products fail to reach the intended audience, special emphasis is placed on disseminating TCRP results to the intended end users of the research: transit agencies, service providers, and suppliers. TRB provides a series of research reports, syntheses of transit practice, and other supporting material developed by TCRP research. APTA will arrange for workshops, training aids, field visits, and other activities to ensure that results are implemented by urban and rural transit industry practitioners.

The TCRP provides a forum where transit agencies can cooperatively address common operational problems. The TCRP results support and complement other ongoing transit research and training programs.

TCRP REPORT 113

Price $34.00

Project H-28
ISSN 1073-4872
ISBN-10: 0-309-09861-0
Library of Congress Control Number 2006906799

© 2006 Transportation Research Board

COPYRIGHT PERMISSION

Authors herein are responsible for the authenticity of their materials and for obtaining written permissions from publishers or persons who own the copyright to any previously published or copyrighted material used herein.

Cooperative Research Programs (CRP) grants permission to reproduce material in this publication for classroom and not-for-profit purposes. Permission is given with the understanding that none of the material will be used to imply TRB, AASHTO, FAA, FHWA, FMCSA, FTA, or Transit Development Corporation endorsement of a particular product, method, or practice. It is expected that those reproducing the material in this document for educational and not-for-profit uses will give appropriate acknowledgment of the source of any reprinted or reproduced material. For other uses of the material, request permission from CRP.

NOTICE

The project that is the subject of this report was a part of the Transit Cooperative Research Program conducted by the Transportation Research Board with the approval of the Governing Board of the National Research Council. Such approval reflects the Governing Board's judgment that the project concerned is appropriate with respect to both the purposes and resources of the National Research Council.

The members of the technical advisory panel selected to monitor this project and to review this report were chosen for recognized scholarly competence and with due consideration for the balance of disciplines appropriate to the project. The opinions and conclusions expressed or implied are those of the research agency that performed the research, and while they have been accepted as appropriate by the technical panel, they are not necessarily those of the Transportation Research Board, the National Research Council, the Transit Development Corporation, or the Federal Transit Administration of the U.S. Department of Transportation.

Each report is reviewed and accepted for publication by the technical panel according to procedures established and monitored by the Transportation Research Board Executive Committee and the Governing Board of the National Research Council.

The Transportation Research Board of the National Academies, the National Research Council, the Transit Development Corporation, and the Federal Transit Administration (sponsor of the Transit Cooperative Research Program) do not endorse products or manufacturers. Trade or manufacturers' names appear herein solely because they are considered essential to the clarity and completeness of the project reporting.

Published reports of the

TRANSIT COOPERATIVE RESEARCH PROGRAM

are available from:

Transportation Research Board
Business Office
500 Fifth Street, NW
Washington, DC 20001

and can be ordered through the Internet at
http://www.national-academies.org/trb/bookstore

Printed in the United States of America

THE NATIONAL ACADEMIES
Advisers to the Nation on Science, Engineering, and Medicine

The **National Academy of Sciences** is a private, nonprofit, self-perpetuating society of distinguished scholars engaged in scientific and engineering research, dedicated to the furtherance of science and technology and to their use for the general welfare. On the authority of the charter granted to it by the Congress in 1863, the Academy has a mandate that requires it to advise the federal government on scientific and technical matters. Dr. Ralph J. Cicerone is president of the National Academy of Sciences.

The **National Academy of Engineering** was established in 1964, under the charter of the National Academy of Sciences, as a parallel organization of outstanding engineers. It is autonomous in its administration and in the selection of its members, sharing with the National Academy of Sciences the responsibility for advising the federal government. The National Academy of Engineering also sponsors engineering programs aimed at meeting national needs, encourages education and research, and recognizes the superior achievements of engineers. Dr. William A. Wulf is president of the National Academy of Engineering.

The **Institute of Medicine** was established in 1970 by the National Academy of Sciences to secure the services of eminent members of appropriate professions in the examination of policy matters pertaining to the health of the public. The Institute acts under the responsibility given to the National Academy of Sciences by its congressional charter to be an adviser to the federal government and, on its own initiative, to identify issues of medical care, research, and education. Dr. Harvey V. Fineberg is president of the Institute of Medicine.

The **National Research Council** was organized by the National Academy of Sciences in 1916 to associate the broad community of science and technology with the Academy's purposes of furthering knowledge and advising the federal government. Functioning in accordance with general policies determined by the Academy, the Council has become the principal operating agency of both the National Academy of Sciences and the National Academy of Engineering in providing services to the government, the public, and the scientific and engineering communities. The Council is administered jointly by both the Academies and the Institute of Medicine. Dr. Ralph J. Cicerone and Dr. William A. Wulf are chair and vice chair, respectively, of the National Research Council.

The **Transportation Research Board** is a division of the National Research Council, which serves the National Academy of Sciences and the National Academy of Engineering. The Board's mission is to promote innovation and progress in transportation through research. In an objective and interdisciplinary setting, the Board facilitates the sharing of information on transportation practice and policy by researchers and practitioners; stimulates research and offers research management services that promote technical excellence; provides expert advice on transportation policy and programs; and disseminates research results broadly and encourages their implementation. The Board's varied activities annually engage more than 5,000 engineers, scientists, and other transportation researchers and practitioners from the public and private sectors and academia, all of whom contribute their expertise in the public interest. The program is supported by state transportation departments, federal agencies including the component administrations of the U.S. Department of Transportation, and other organizations and individuals interested in the development of transportation. **www.TRB.org**

www.national-academies.org

COOPERATIVE RESEARCH PROGRAMS

CRP STAFF FOR TCRP REPORT 113

Robert J. Reilly, *Director, Cooperative Research Programs*
Christopher W. Jenks, *TCRP Manager*
S. A. Parker, *Senior Program Officer*
Eileen P. Delaney, *Director of Publications*
Natalie Barnes, *Editor*

TCRP PROJECT H-28 PANEL
Field of Policy and Planning

James W. Kemp, *New Jersey Transit Corporation, Newark, NJ* (Chair)
Fabian Cevallos, *University of South Florida, Weston, FL*
Thomas Friedman, *King County (WA) Metro Transit, Seattle, WA*
Erin Mitchell, *Metro Transit—Minneapolis/St. Paul, Minneapolis, MN*
Yuko Nakanishi, *Nakanishi Research and Consulting, Rego Park, NY*
Gerald Pachucki, *Utah Transit Authority, Salt Lake City, UT*
Kimberly Slaughter, *S.R. Beard & Associates, LLC, Houston, TX*
Wei-Bin Zhang, *University of California—Berkeley, Richmond, CA*
Sarah Clements, *FTA Liaison*
Louis F. Sanders, *APTA Liaison*
Kay Drucker, *Bureau of Transportation Statistics Liaison*
Peter Shaw, *TRB Liaison*

FOREWORD

By S. A. Parker
Staff Officer
Transportation Research Board

This report will be of interest to transit personnel responsible for planning, scheduling, and managing reliable bus transit services in congested areas. This report will also be useful to other members of technology procurement teams, representing operations, maintenance, information systems, human resources, legal, finance, and training departments.

In response to growing traffic congestion and passenger demands for more reliable service, many transit operators are seeking to improve bus operations by investing in automatic vehicle location (AVL) technology. In addition, automatic passenger counters (APCs), which can collect passenger-activity data compatible with AVL operating data, are beginning to reach the mainstream. Many operators are planning, implementing, or operating AVL-APC systems. The primary application of AVL technology has been in the area of real-time operations monitoring and control; consequently, AVL data has not typically been stored or processed in a way that makes it suitable for subsequent, off-line analysis. In contrast, APC data is generally accessed for reporting and planning purposes long after opportunities for real-time use have expired.

Beyond the area of real-time operations control, AVL technology holds substantial promise for improving service planning, scheduling, and performance analysis practices. These activities have historically been hampered by the high cost of collecting operating and passenger-activity data; however, AVL and APC systems can capture the large samples of operating data required for performance analysis and management at a fairly low incremental cost. Compared to real-time applications of AVL data, off-line analysis of archived data has different demands for accuracy, detail, and ability to integrate with other data sources. Operators and vendors need effective strategies for designing AVL-APC systems to capture and process data of a quality needed for off-line analysis, and for archiving and taking advantage of this promising data source.

The objective of TCRP Project H-28 was to develop guidance for the effective collection and use of archived AVL-APC data to improve the performance and management of transit systems. This report offers guidance on five subjects:

- Analyses that use AVL-APC data to improve management and performance
- AVL-APC system design to facilitate the capture of data with the accuracy and detail needed for off-line data analysis
- Data structures and analysis software for facilitating analysis of AVL-APC data
- Screening, parsing, and balancing automatic passenger counts
- Use of APC systems for estimating passenger-miles for National Transit Database reporting

Tools for (1) analyzing running times and (2) designing scheduled running times using archived AVL and APC data were created as an extension of the existing software TriTAPT (Trip Time Analysis in Public Transport), a product of the Delft University of Technology. In addition, TriTAPT was used to demonstrate one of the advanced data structures recommended, that of a "virtual route" consisting of multiple overlapping routes serving the same street. Under the terms of this project, TriTAPT is available without license fee to U.S. and Canadian transit agencies through 2009. To request TriTAPT, please send an email to TriTAPT@neu.edu.

From the TRB website (http://www4.trb.org/trb/crp.nsf/All+Projects/TCRP+H-28), the following items can be accessed: (1) an electronic version of this report; (2) spreadsheet files with prototype analyses of passenger waiting time (using AVL data) and passenger crowding (using APC data); and (3) case studies (as appendixes to *TCRP Web Document 23: Uses of Archived AVL-APC Data to Improve Transit Performance and Management: Review and Potential*).

CONTENTS

- **1** Summary
- **8** **Chapter 1** Introduction
  - 8   1.1 Historical Background
  - 9   1.2 Research Objective
  - 10   1.3 Research Approach
  - 12   1.4 Report Outline
- **14** **Chapter 2** Automatic Vehicle Location
  - 14   2.1 Location Technology
  - 14   2.2 Route and Schedule Matching
  - 17   2.3 Data Recording: On- or Off-Vehicle
  - 19   2.4 Data Recovery and Sample Size
- **21** **Chapter 3** Integrating Other Devices
  - 21   3.1 Automatic Passenger Counters
  - 21   3.2 Odometer (Transmission Sensors)
  - 22   3.3 Door Switch
  - 22   3.4 Fare Collection Devices
  - 22   3.5 Other Devices
  - 23   3.6 Integration and Standards
- **25** **Chapter 4** Uses of AVL-APC Data
  - 25   4.1 Becoming Data Rich: A Revolution in Management Tools
  - 29   4.2 Key Dimensions of Data Needs
  - 29   4.3 Targeted Investigations
  - 29   4.4 Running Time
  - 33   4.5 Schedule Adherence, Long-Headway Waiting, and Connection Protection
  - 35   4.6 Headway Regularity and Short-Headway Waiting
  - 36   4.7 Demand Analysis
  - 39   4.8 Mapping
  - 39   4.9 Miscellaneous Operations Analyses
  - 40   4.10 Higher Level Analyses
- **41** **Chapter 5** Tools for Scheduling Running Time
  - 41   5.1 Running Time Periods and Scheduled Running Time
  - 43   5.2 Determining Running Time Profiles Using the Passing Moments Method
- **45** **Chapter 6** Tools for Analyzing Waiting Time
  - 45   6.1 A Framework for Analyzing Waiting Time
  - 46   6.2 Short-Headway Waiting Time Analysis
  - 48   6.3 Long-Headway Waiting Time Analysis

51 Chapter 7 Tools for Analyzing Crowding
- 51 7.1 Distribution of Crowding by Bus Trip
- 51 7.2 Distribution of Crowding Experience by Passenger

54 Chapter 8 Passenger Count Processing and Accuracy
- 54 8.1 Raw Count Accuracy
- 55 8.2 Trip-Level Parsing
- 58 8.3 Trip-Level Balancing Methods

63 Chapter 9 APC Sampling Needs and National Transit Database Passenger-Miles Estimates
- 63 9.1 Sample Size and Fleet Penetration Needed for Load Monitoring
- 63 9.2 Accuracy and Sample Size Needed for Passenger-Miles

66 Chapter 10 Designing AVL Systems for Archived Data Analysis
- 66 10.1 Off-Vehicle versus On-Vehicle Data Recording
- 66 10.2 Level of Spatial Detail
- 68 10.3 Devices to Include
- 69 10.4 Fleet Penetration and Sampling
- 69 10.5 Exception Reporting versus Exception Recording

70 Chapter 11 Data Structures That Facilitate Analysis
- 70 11.1 Analysis Software Sources
- 72 11.2 Data Screening and Matching
- 73 11.3 Associating Event Data with Stop/Timepoint Data
- 74 11.4 Aggregation Independent of Sequence
- 75 11.5 Data Structures for Analysis of Shared-Route Trunks
- 75 11.6 Modularity and Standard Database Formats

77 Chapter 12 Organizational Issues
- 77 12.1 Raising the Profile of Archived Data
- 77 12.2 Management Practices to Support Data Quality
- 77 12.3 Staffing and Skill Needs
- 78 12.4 Managing an Instrumented Sub-fleet
- 78 12.5 Avoiding Labor Opposition

79 Chapter 13 Conclusions

81 References

83 Appendixes

SUMMARY

Using Archived AVL-APC Data to Improve Transit Performance and Management

Automatic vehicle location (AVL) and automatic passenger counter (APC) systems are capable of gathering an enormous quantity and variety of operational, spatial, and temporal data that, if captured, archived, and analyzed properly, holds substantial promise for improving transit performance by supporting improved management practices in areas such as service planning, scheduling, and service quality monitoring. Historically, however, such data has not been used to its full potential. Many AVL systems, designed primarily for real-time applications, fail to capture and/or archive data items that would be valuable for off-line analysis. And where good quality data is captured, new analysis tools are needed that take advantage of this resource.

Recent technological advances have created new opportunities for improving the quantity, variety, and quality of data captured and for analyzing it in meaningful ways. The objective of this research was to develop guidance for the effective collection, archiving, and use of AVL-APC data to improve the performance and management of transit systems.

This project yielded three types of products: a survey of practice, guidance on AVL-APC systems and data analysis, and prototype tools for analysis of archived AVL-APC data. The state of the practice in AVL-APC data capture and analysis was ascertained by means of literature review, widespread telephone interviews, intensive case studies of nine transit agencies in three countries, and a workshop for suppliers. The case studies (published as *TCRP Web Document 23*) are from five transit agencies in the United States (Seattle [WA], Portland [OR], Chicago [IL], New Jersey, and Minneapolis [MN]); two agencies in Canada (Ottawa, Montreal); and two agencies in the Netherlands (The Hague and Eindhoven).

This report offers guidance on five subjects:

- Analyses that can use AVL-APC data to improve management and performance
- AVL-APC system design that facilitates the capture of data with the accuracy and detail needed for off-line data analysis
- Data structures and analysis software for facilitating analysis of AVL-APC data
- Screening, parsing, and balancing automatic passenger counts
- Use of APC systems for estimating passenger-miles for National Transit Database (NTD) reporting

The tools developed for analyzing AVL and APC data are described in this report; in addition, provision has been made for their distribution. Prototype analyses of passenger waiting time (using AVL data) and passenger crowding (using APC data), developed on a spreadsheet platform, are available from the project description web page for TCRP Project H-28 on the TRB website (www.trb.org). Tools for analyzing running times and designing scheduled running times were created as an extension of the already existing software TriTAPT (Trip Time Analysis in Public Transport), a product of the Delft University of Technology. TriTAPT was also used to demonstrate one of the advanced data structures recommended, that of a "virtual route" consisting of multiple overlapping routes serving the same street. Under the terms of this project, TriTAPT is available without license fee to U.S. and Canadian transit agencies for 4 years.

AVL and APC System Design

For their primary use, AVL systems include a reliable means of location and APC systems include sensors and algorithms that count passengers entering and exiting. However, archived data analysis requires data beyond what is needed for the primary purpose of AVL system design, real-time monitoring.

Most important for archived data analysis is the ability to match raw location data to a base map and schedule. Inability to match data is the primary cause for rejecting data from database archives; data recovery rates as poor as 25% to 75% have been reported for APCs, although rates are generally far better for AVL. Success in matching depends to a large extent on the data captured. If the AVL system is integrated with a radio, operator sign-in including route-run number can be captured, which aids matching. Systems that detect door openings and correlate them with stops have an advantage over those without door sensors, because stops are natural match points. The most difficult part of matching is correctly identifying the end of the line, because vehicle behavior at the end of a line is less predictable. Better matching algorithms are needed especially for end-of-line detection. With detailed information about vehicle movements and door openings in the terminal area, algorithms can better determine when a bus arrives at the end of a line and when a bus begins a trip.

AVL systems produce three kinds of frequent records: polling records, stop records, and timepoint records. Polling records indicate a bus's location when queried by a central computer doing round-robin polling. Polling is intended mainly for real-time knowledge of vehicle location; its records can be called "location-at-time" data as opposed to stop and timepoint records, which are "time-at-location" data. Most analyses of AVL-APC data need time at specific locations, for example, to analyze running time or schedule adherence. While estimation of time-at-location is possible by interpolation from polling data, it has so far proven impractical. Polling data's only value for off-line analysis is for incident investigation, using "playback."

Stop and timepoint records include the time at which the bus departed and/or arrived at specified points. Different system designs define the time being recorded differently; for example, arrival time may be when the bus enters within a zone of a 10-m radius about a stop or when the first door opens (or the bus passes the stop if the door is not opened). Door switch and odometer connections enable more precise sensing of arrival and departure times, improving data accuracy. Because some analyses use arrival time while others use departure time, recording both increases the usefulness of the data archive.

Stop records offer greater geographic detail than timepoint records. On buses with APC, records are always made at the stop level; with AVL, timepoint records are more common. However, there are advantages to making stop records even when there are no passenger counts. The advantages include the ability to detect holding (i.e., a bus remaining at a stop with doors closed or with doors open but for an unusually long time, as when a bus is ahead of schedule); geographic detail for analyzing delays; and the ability to make stop-level schedules, whether for publishing, for computer-based trip planners, or for next bus arrival prediction systems.

AVL-APC systems also make types of records other than these basic, frequent records. Radio-based systems create event records, sometimes for more than 100 event types, all stamped with time and location. They can include events that are generated automatically (e.g., engine turned on or off) and events that are operator initiated and sent by data radio (e.g., pass-up, railroad crossing delay). Event records useful for data analysis include pass-ups, various types of delay (e.g., drawbridge, railroad crossing), indications of special user types (e.g., wheelchair lift users, bicycle rack users), and events that help with matching the trip. If the on-board computer monitors bus speed and heading, records can be written for noteworthy changes (e.g., when speed crosses a crawl speed threshold) or for the maximum speed achieved between stops.

AVL systems can also be configured to make records very frequently (e.g., every 2 s). That kind of data is helpful for mapping trajectories and for capturing speed-related details.

AVL systems have two options for data recording: (1) via an on-board computer that is uploaded nightly, usually using an automated, high-speed link, or (2) via real-time radio transmission, with records stored in the receiving computer. Because data radio has limited transmission capacity, this mode limits the frequency (and to a lesser extent, length) of records that can be made. On-board storage, in contrast, presents no practical capacity limitation and is therefore inherently better suited to data collection.

On-board devices that can be integrated with an AVL system include APCs, radio control head, odometer (transmission), gyroscope, door sensors, wheelchair lift sensor, farebox, and stop enunciator. In general, the more devices included, the richer the data stream, which can aid both in matching and in offering new kinds of information. Integration with the radio control head enables the system to record operator-initiated messages as events and to capture valuable sign-in information. Integration with the odometer provides a backup to the geographic positioning system (GPS) and information on speed and can be used to detect when a bus starts and stops moving, which aids in determining arrival and departure times and delay between stops. Door sensors are valuable for detecting arrival and departure times at stops, as well as for matching. Integration with a newer farebox or other fare collection system that makes transaction records offers the possibility of getting location-stamped farebox records, which can be a valuable source of ridership data, especially where APCs are lacking.

Integration with a stop announcement system does not add any new data to the system, but because stop announcements demand careful matching, it helps ensure that the location and matching system has high reliability. Likewise, integration with real-time radio offers the possibility of detecting and correcting, with operator or dispatcher assistance, errors such as invalid sign-in information (e.g., a run number that does not agree with where a bus is found to be operating).

Analyses Using Archived AVL-APC Data

A large number of uses for archived AVL-APC data were identified, and their various needs for record type and data detail were analyzed.

Targeted investigations apply to passenger complaints, legal claims, and payroll disputes, among others. They require only the ability to, in effect, play back a bus's trajectory and, therefore, can be done with polling data. The greater the detail of the data stream, including the frequency of records, the more its potential uses for targeted investigations.

One of the richest application areas for archived AVL data is in *running time analysis,* including designing scheduled running times and monitoring schedule adherence. Traditional scheduling methods, created in an era of expensive manual data collection, are based on mean observed running times, which are estimated from small sample sizes. Now, however, AVL data offers the possibility of using extreme values such as 85- and 95-percentile running times as a basis for scheduling. Extreme values are important to passengers, who care less about mean schedule deviation than about avoiding extreme deviations such as buses that are early or very late. Extreme values are also important to planning, because a goal of cycle time selection (sum of scheduled running time and scheduled layover) is to limit the probability that a bus finishes one trip so late that its next trip starts late. For transit agencies that use holding to prevent early departures, data analysis can try to identify holding time, allowing the analysis of net running time for making schedules. Most running time analyses can be performed with timepoint-level data, but there are advantages to using stop records, as mentioned earlier.

As transit agencies take a more active role in improving operating speed and protecting their routes from congestion, analyses of speed and delay along a route become valuable tools. Therefore, stop-level detail is important for determining where delays are occurring and monitoring the effects of local changes to traffic conditions.

AVL data also can be applied to *schedule adherence, headway regularity, and passenger waiting time*. In this arena, too, extreme values are at least as important as mean values. Also, because extreme values can only be estimated reliably from a large sample size, the availability of archived AVL data offers new opportunities for analysis. In particular, transit agencies are looking for measures of service quality that reflect passengers' experience and viewpoint. Tools developed by this project for measuring waiting time, service reliability impacts, and crowding demonstrate this possibility.

APC data lends itself to a variety of passenger *demand analyses* including determining load profiles and using demand rates to set headways and departure times. While traditional analyses rely on mean values, APC data also offers the possibility of focusing on extreme values of crowding, as well as relatively rare events such as wheelchair lift and bicycle rack use.

AVL data containing stop records can be used to verify and update base maps. By analyzing where buses actually stop, one can update stop locations. Some AVL systems deliberately include a "learning mode" in which bus location is recorded frequently enough to give the bus's path where route information is needed for the base map, such as through a shopping center or a new subdivision.

Exploring archived AVL-APC data can enable transit agencies to find hidden trends that help explain irregularities in *operations* and suggest new avenues for improvement. As an example, one agency found that a surprising amount of schedule deviation could be explained by the operator—that is, some operators consistently depart late or run slow—which suggests the need for improved methods of operator training and supervision.

An example of an advanced analysis using a highly detailed AVL data stream is calculating and monitoring measures of ride smoothness. A smooth ride is certainly important to passengers, and AVL data with either very frequent observations or accelerometers can allow ride smoothness to be measured objectively.

For any of the detailed analyses mentioned, there is also interest in *higher level analysis* involving tracking trends over time, comparing routes or periods of time, and so forth. Another example is geographic information system (GIS)-based analyses that take as input passenger use and service quality statistics.

Prototype Analysis Tools: Waiting Time and Crowding

Analysis tools were developed for passenger waiting time on short- and long-headway services, for crowding, and for designing scheduled running time. The tools developed for waiting time analysis use some newly proposed measures that reflect the amount of time passengers budget for waiting, which is particularly sensitive to service reliability. These proposed measures are based on extreme values of the headway and schedule deviation distribution, which can only be estimated using the large sample sizes that AVL datasets provide.

For short-headway service, for which passengers can be assumed to arrive independent of the schedule, the distribution of passenger waiting time can be determined from headway data. The tools developed include graphing the distribution of waiting time, determining mean and 95-percentile waiting time, and determining the percentage of passengers whose waiting time falls into various user-defined ranges. The latter is useful for supporting a service quality standard such as "no more than 5% of passengers should have to wait more than 2 minutes longer than the scheduled headway."

For short-headway service, "budgeted waiting time" is taken to be the 95-percentile waiting time. The difference between budgeted waiting time and the mean time passengers actually spend waiting is called "potential waiting time." While potential waiting time is not spent waiting on the platform, it still involves a real cost to passengers. "Equivalent waiting time" is a proposed measure of passenger waiting cost, being a weighted sum of platform waiting time (with proposed weight = 1) and potential waiting time (with proposed weight = 0.5). Example analyses and reports illustrate the concept and show how improved headway regularity reduces passenger waiting cost.

Waiting time and its components also can be divided between a part that is "ideal" (i.e., what it would be if service exactly followed the schedule) and the remainder ("excess" waiting time). This division separates the effects of planning and operations on passenger waiting.

For long-headway service, excess platform waiting time is defined as the difference between the mean and the 2-percentile departure deviation, based on the idea that experienced passengers will arrive early enough to limit to 2% or less their probability of missing the bus. Potential waiting time is the difference between the 95-percentile and mean departure deviation, and excess equivalent waiting time is a weighted sum of these components. Example analyses and reports show how these waiting time measures are sensitive to improvements in service reliability (in this case, on-time performance).

Traditional measures of crowding are mean maximum load and the percentage of trips in various crowding ranges. Neither reflects well the impact of crowding on passengers. With APC data, it is possible to determine the number of passengers experiencing various levels of crowding ranging from "seated next to an unoccupied seat" to "standing at an unacceptable level of crowding."

The analysis tools used for waiting time and crowding are included on a spreadsheet platform on the project description web page for TCRP Project H-28 on the TRB website (www.trb.org).

Prototype Analysis Tools: Scheduled Running Time

Analysis tools were also developed for running time. One is an interactive tool for determining scheduled running times (allowed times) across the day, including selecting boundaries between periods of homogeneous running times. This tool includes an automated portion, in which boundaries and allowed times are selected based on user-supplied criteria regarding feasibility (probability that trips can be completed in their allowed time) and tolerance. On a graphical interface, users can modify both period boundaries and allowed times by simple drag-and-drop; the program will respond immediately with the feasibility of the proposed changes.

The second main tool is for allocating running time along the route, thereby determining running time on each segment. It uses the Passing Moments method of maintaining a given probability of completing a trip on time, in order to give operators an incentive to hold when they are ahead of schedule, thus improving schedule adherence.

The analysis tools for running time are part of the TriTAPT package, which is being distributed without license fee to U.S. and Canadian transit agencies as part of this project.

Processing and Using Automatic Passenger Counts

Accuracy of passenger counts is always a concern with automated systems. Analysis of the various dimensions of accuracy confirmed theoretical findings with published findings of APC data accuracy. The report shows that systematic under- or overcounting is a more serious problem than random errors, because of the large sample size afforded by APCs. The accuracy of load and passenger-miles measures can be substantially worse than the accuracy of on and off counts because of the way load calculations allow errors to accumulate.

Getting accurate load and passenger-miles estimates from automatic passenger counts demands not only relatively accurate counts, but also good methods of parsing the data into trips and balancing on-off discrepancies. Therefore, checking the accuracy of APC-measured load against manual counts is a good system test.

To prevent drift in load estimates, a day's data stream of automated counts has to be parsed at points of known load. Those points are usually layover points; therefore, parsing usually means dividing the data into trips. This need requires the end of the line to be clearly defined for APC systems.

Failing to parse data at the trip level can bias load and passenger-miles estimates upwards because downward errors cause negative loads, which are routinely corrected when any trip's data is processed, while upward errors are not readily apparent except at the block (vehicle assignment) level.

Where load does not necessarily become zero at the end of a trip, whether due to a route ending in a loop or to through-routing, data structures must account for passengers inherited from a previous trip, either by adding dummy stop records or using a field in a trip header record.

Algorithms are presented for parsing data for trips that end with short loops, making the assumptions that (1) nobody rides all the way around the loop and (2) nobody's trip lies entirely within the loop. Within the loop, passengers alighting are attributed to the trip entering the loop, and passengers boarding are attributed to the trip leaving the loop. With these assumptions, ons and offs can be balanced just as if load were known to be zero at the end of the trip, even if actual load on the bus never goes to zero. If a stop within the loop is designated as the route endpoint, inherited passengers can be determined at that point.

An algorithm for balancing ons and offs at the trip level is presented. It includes calculation of the most likely value of total ons and offs, proportional corrections to stop-level counts, and rounding to integer values. It accounts for inherited passengers and checks not only against negative departing load, but also against negative through load, which is often a tighter condition. It also makes a first attempt to account for operator movements off and on the bus.

Because APCs are normally deployed on only a percentage of the fleet that is rotated around the system, sample size can vary substantially between routes and trips. Data analysis using APC data should account for varying sample size, weighting not each observation, but each operated trip, equally. Most uses of passenger counts require only moderate sample sizes, which can be met with the typical 10% to 15% fleet penetration. The only exception is monitoring crowding levels on crowded routes; because of the importance of estimating extreme values of crowding, that application requires a large sample size.

For NTD reporting, estimates of passenger-miles made from APC data can easily meet the specified accuracy level, provided systematic under- or overcounts are limited. The report shows the relationship between required sample size and bias and shows that, for all but the smallest transit systems, NTD accuracy can be obtained even if only a small percentage of the fleet is instrumented with APCs.

Data Structures That Facilitate Analysis

Many analyses are tied to a route pattern, that is, a sequence of stops. Examples are load profiles, headway irregularity, and running time. For such analyses, the basic unit is the stop or timepoint record. Data structures are needed to indicate the sequence of stops and the schedule.

When the data stream includes information on events that happened on segments between stops or timepoints, that information can be used in a pattern-based analysis only if it can be associated with a stop or timepoint. One approach is to include a summary of segment information (e.g., total delay on the segment or maximum speed on the segment) in fields in the stop or timepoint records. An alternative and more flexible approach is to include in each event record a field for the stop with which it can be associated.

Header records for trips, blocks, and days can speed analysis by allowing selection filters to be applied to a much smaller number of records. Making summary records at the trip level—containing such summary measures for a trip as total ons and offs, maximum load, passenger-miles, and running time—will speed analysis of the standard measures that are summarized. Summary records also can be created over standard date ranges (e.g., every month or quarter) for higher level analysis, such as trends or historical comparisons. However, some agencies find that their database systems are able to generate summaries on the fly quickly enough that summary records are not needed.

Transit agencies often want analyses to cover multiple route patterns (e.g., all the patterns that compose a line or all the patterns operating along a certain corridor or in a certain geographic area). If the analysis does not relate to a particular sequence of stops, the data can be analyzed by simple aggregation; for example, total boardings or total number of early departures can simply be aggregated over any group of stops. However, an analysis that involves relating the data to a sequence of stops used by multiple patterns—such as running time, headway, and load on a trunk—requires a data structure specifying the sequence of stops. A data structure called a "virtual route" was developed and implemented in TriTAPT as a proof of concept. It allows users to do almost any analysis on all the patterns that serve, in part or whole, a given sequence of stops.

The development of software tools for analyzing AVL-APC data is still in its infancy. There are advantages both to solutions customized by or for individual transit agencies and to solutions that are common to multiple transit agencies. Customized solutions have worked well for some transit agencies, but require a level of database programming expertise unavailable to many transit agencies. Market mechanisms for shared solutions—which offer the possibility of greater expertise, economy, and continual upgrading (a necessity with today's information technology)—include AVL and APC equipment suppliers, scheduling software suppliers, and third-party software suppliers.

Ideally, externally supplied software should be modular with respect to the native data format and ultimate report formats. It should provide an open interface for data input, so that it is not restricted to a single type of data collection system. While it may be programmed to deliver various types of reports, it also should offer the possibility of exporting tables as a result of its analysis, giving agencies the freedom to format as they wish. In addition, transit agencies should always maintain the freedom to manipulate and explore their archived data beyond the tools provided by any externally supplied software.

CHAPTER 1

Introduction

AVL and APC systems are capable of gathering an enormous quantity and variety of operational, spatial, and temporal data that—if captured, archived, and analyzed properly—hold substantial promise for improving transit performance by supporting improved management practices in areas such as service planning, scheduling, and service quality monitoring. Historically, however, such data has not been used to its full potential. Many AVL systems, designed primarily for real-time applications, fail to capture and/or archive data items that would be valuable for off-line analysis. Recent technological advances have created new opportunities for improving the quantity, variety, and quality of captured and archived data and for analyzing it in meaningful ways. The objective of this research was to develop guidance for the effective collection, archiving, and use of AVL-APC data to improve the performance and management of transit systems.

Automatically collected data can play two important roles in a transit agency's service quality improvement process. As illustrated in Figure 1, there are two quality improvement cycles: one in real time and one off line. In the real-time loop, automatically collected data drives operational control, aiding the transit agency in detecting and responding to deviations from the operational plan; it is also a source of real-time information that can be conveyed to customers using a variety of media. In the off-line loop, automatically collected data that has been archived drives analyses that aid the transit agency in evaluating and improving its operational plan. Ultimately, good operational performance and high passenger satisfaction follow from having both a good operational plan and good operational control.

1.1 Historical Background

Historically, AVL system design has emphasized the real-time loop, giving little or no attention to the off-line loop. Many AVL systems do not archive data in a manner useful for off-line analysis because they were not designed to do so. The inability of AVL to deliver data for off-line planning analysis was a major theme of a 1988 conference sponsored by the Canadian Urban Transit Association [see, for example, Furth (*1*)]. Through the mid-1990s, this situation continued. A report summarizing data analysis practice in 1998 found that, of the seven U.S. transit agencies surveyed that had AVL systems, all relied entirely on manual data collection for running time analysis, and only three used AVL data to monitor schedule adherence (*2*).

Broward County Transit illustrates how AVL systems are not commonly oriented toward archiving data (*3*). Its AVL system archived incident messages, but not routine poll data, which gave vehicle location approximately every 60 s. Because incident messages occur only on an exception basis, they cannot support most running time and schedule adherence analyses. Although the poll messages were written to an Oracle database as a utility for system maintainers, it was overwritten every 2 min because no permanent use for that data was foreseen. To archive the poll data, an analyst wrote a program that copies the contents of the Oracle database to a permanent database every 2 min. This archive enabled him to plot trajectories and review particular trips.

Because real-time data needs differ from those of off-line analysis, simply warehousing AVL records does not in itself guarantee a useful data archive. The single greatest problem with traditional AVL data is that it consists mostly of poll records, in which a vehicle reports its location when polled by a central computer in round-robin fashion, every 60 seconds or so. Poll data can be characterized as "location-at-time" data, as distinct from "time-at-location" data such as reporting when a bus arrives at a stop. Either one is adequate for tracking a bus's location in real time, but most off-line analyses (e.g., running time, headway, or schedule adherence) need records of when buses arrive or depart from stops and timepoints.

There are other important differences between real-time and archived data needs. Some AVL systems transmit data only when a bus's schedule deviation is outside a "normal" range

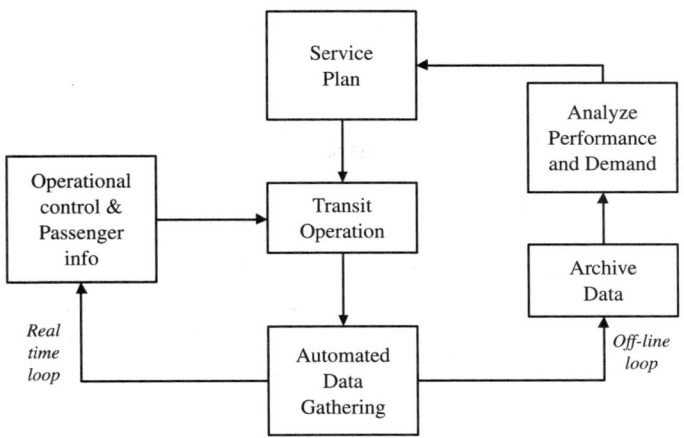

Figure 1. Service quality improvement cycles.

(e.g., 0 to 3 min late). For real-time control, or even passenger information, exception data like this may be adequate; however, a data stream containing only exception data is entirely unsuitable for analyzing running time. Another difference is that real-time data is more tolerant of errors. If faults in either the location system or the base map make a bus's apparent track momentarily jump a few miles off route, service controllers learn quickly how to ignore such anomalies; however, in a data archive, such errors may go undetected and distort running time or schedule deviation analyses.

APC systems, unlike AVL, have always been oriented toward off-line analysis. Historically, APC systems were independent of (real-time) AVL systems. Their adoption has been far more limited than AVL systems. The weaker market for APCs has resulted in several vendors having gone out of business and has limited the amount of vendor-developed software for data analysis. (That trend, fortunately, is being reversed as technology advances.) Transit agencies have been largely left on their own to develop their APC systems; pioneering successes include transit systems in Seattle, Ottawa, Winnipeg, and Toronto.

A radical shift in AVL system development in favor of archived data collection occurred in the mid-1990s, when Tri-Met, together with an AVL vendor, designed a hybrid AVL-APC system featuring on-board event recording *and* radio-based communication. Records for every stop and other events are created in the on-board computer, while messages useful for real-time monitoring (e.g., bus is late or off-route) are radioed in. Another innovation was that APCs, installed on a sample of the fleet, do not have their own location system; they simply count passengers and rely on the AVL system to provide the location stamp. This design lowered the marginal cost of instrumenting buses with APCs to the point that all new buses at Tri-Met now have APCs. With 65% of the fleet equipped, Tri-Met stands out as the only agency with more than 20% of its fleet APC-equipped. After a few years' break-in period, Tri-Met enjoys a comprehensive archive of high-quality location data, making it a leader in archived AVL-APC data analysis.

NJ Transit took a still more radical step in designing, in conjunction with an AVL vendor, an AVL-APC system that, for the time being at least, does not include real-time radio communication. Although they envision integrating radio in the future, the current configuration is strictly for off-line analysis.

In the Netherlands, AVL systems with on-board data recording were installed in some cities in the early 1990s. However, the lack of analysis tools and poor quality of base maps led to their data being largely unused until about 1996, when the Delft University of Technology teamed up with the municipality of Eindhoven and its transit operator, Hermes, to improve their data quality and develop and apply analysis tools. Those tools, originally developed around 1980 for use with manually collected data, were coded into the software package TriTAPT (Trip Time Analysis in Public Transport), which has also been applied at transit agencies of the Hague, Rotterdam, and Utrecht.

A significant recent development for archived AVL-APC data is the growing demand for stop announcement systems, which require considerable on-board computing power to match stops in real time. The creators of those systems found it easy to simply add data recording, creating a new function for their product and thus creating a new source of archived AVL data—one that promises to be very accurate by virtue of being designed for a rather demanding application. In the system installed in the Chicago Transit Authority (CTA), APCs have also been integrated on a sample of the fleet.

The movement to integrate systems and on-board devices is serving to blur the distinction between what first appeared as stand-alone systems. In recent years, for example, some cities, including Minneapolis, have integrated APCs into real-time AVL systems, with stop data transmitted over the air to a central computer for archiving. APC and event recording has been added to stop announcement systems, and stop announcements have been added to real-time AVL systems. However, regardless of what functions a data collection system may have, there are common needs for archived data analysis. Therefore, this report will sometimes use the term AVL in a generic sense, meaning any automatic data collection system that includes location data.

1.2 Research Objective

Technological advances erase all question of the feasibility of collecting and archiving high-quality location data. Two primary issues remain:

- How should automatic data collection systems and their associated databases and software be designed so that they capture and facilitate the analysis of AVL-APC data?

- What new analysis tools might become possible using this kind of archived data?

1.3 Research Approach

The research followed three major thrusts, described in this section:

- Survey of industry practice
- Analysis of data systems and opportunities
- Development of analysis tools

1.3.1 Survey of Practice: Breadth and Depth

Six information sources were used to review (1) AVL, APC, and related systems with respect to their ability to capture and archive operational data and (2) industry practice in using such archived data. Together, these sources provide both a broad view of historical and current practice and an in-depth view of practice at transit agencies in the United States, Canada, and Europe.

The first source was the literature on intelligent systems in transit and on transit analysis tools. A helpful starting point was a pair of state-of-practice reviews done for the U.S.DOT's Volpe Center (4, 5).

A second source was a mail survey of U.S. transit agencies concerning their use of AVL and APC systems conducted in Spring 2001 by Robbie Bain.

A third source was a wide telephone survey of APC and AVL users. From the first two sources, the researchers assembled a preliminary list of some 122 U.S., 14 Canadian, and 26 European transit agencies using or planning to use AVL or APC systems. From that list, staff members were telephoned at transit agencies reputed to be advanced in their use of AVL or APC data. Telephone interviews were successfully conducted with the 20 U.S. and 14 Canadian transit agencies listed in Table 1.

The fourth source was in-depth case studies at nine transit agencies in the United States, Canada, and the Netherlands, listed in Table 2. Appendixes A through I of *TCRP Web Document 23* (available from the TRB website: www.trb.org) contain the case study reports. In addition, a partial case study was conducted of Uestra, the transit agency in Hannover, Germany.

The nine case study agencies provide a broad range in many respects. They represent the United States, Canada, and Europe. Within the United States, they span the East Coast, Midwest, and West Coast and include agencies whose operations are statewide, metropolitan, and limited primarily to a central city. Some have focused on AVL; others on APCs; others on event recorders; and some on two or more of these functions. They represent a range of vendors, system design, and system age. Some of the selected agencies have

Table 1. Agencies interviewed in wide telephone survey.

State/Province	City / Area	Agency
CA	San Jose	Valley Transportation Authority
CO	Denver	Regional Transportation District
FL	Broward County	Broward County Transit
FL	Orlando	LYNX
IA	Des Moines	Metropolitan Transit Authority
IA	Sioux City	Sioux City Transit
IL	Chicago	CTA
IL	Chicago suburbs	Pace
MA	Cape Cod	Cape Code Regional Transit Authority
MD	Baltimore / state	Maryland Transit Administration
MI	Ann Arbor	Ann Arbor Transportation Authority
MN	Minneapolis	Metro Transit
MO	Kansas City	Kansas City Area Transportation Authority
NJ	New Jersey (statewide)	NJ Transit
NY	Buffalo	Niagara Frontier Transportation Authority
OR	Portland	Tri-Met
TX	Dallas	Dallas Area Rapid Transit
TX	San Antonio	VIA
WA	Seattle	King County Metro
WI	Milwaukee	Milwaukee County Transit
AB	Calgary	Calgary Transit
AB	Edmonton	Edmonton Transit
BC	Vancouver	TransLink
BC	Victoria	BC Transit
MB	Winnipeg	Winnipeg Transit
NS	Halifax	Halifax Metro Transit
ON	Hamilton	Hamilton Street Railway
ON	London	London Transport Commission
ON	Ottawa	OC Transpo
ON	Toronto	Toronto Transit Commission
QC	Hull	Société de transport de l'Outaouais
QC	Montreal	STM
QC	Montreal South Shore	Société de Transport de la Rive Sud de Montreal
QC	Quebec	Société de transport de la Communauté urbaine de Quebec

well-established practice with archived data, with impacts throughout the organization; others are still in the development stage. These agencies have had failures as well as successes, and lessons are learned from both.

A brief description of the AVL-APC systems at the case study sites will provide some helpful background.

Table 2. Case study sites.

Location	Agency	Appendix
Portland, OR	Tri-Met	A
New Jersey (statewide)	NJ Transit	B
Seattle, WA	King County Metro	C
Chicago, IL	CTA	D
Montreal, QC	STM	E
Ottawa, ON	OC Transpo	F
Eindhoven, the Netherlands	Hermes	G
The Hague, the Netherlands	HTM	H
Minneapolis, MN	Metro Transit	I

- Tri-Met's system features on-board computers on all of their buses, connected by radio to provide real-time AVL, and stores on-board records for every stop as well as other events. APCs are on about 65% of the fleet. Stop records include on and off counts (void for buses without APCs), stop ID, longitude, latitude, door open moment, dwell (i.e., door open) duration, moment of exiting a 30-m radius zone around the stop, indicators of door opening and lift use, and maximum speed since the previous stop. Location and status are radioed to the control center on an exception basis (e.g., when more than a predetermined deviation from schedule occurs or when the bus is off route). Operator-initiated coded radio messages (e.g., "road blocked by train" or "pass-up") are recorded in the on-board computer with time and location stamp as well as transmitted in real time. The on-board computer is also connected to a traffic signal priority request emitter, triggered only when the bus is behind schedule.
- The HTM (the Hague) system is like Tri-Met's, featuring on-board event recording at every stop, APCs on a fraction of the fleet (in this case, about 25%), radio transmission of real-time location to central control, and traffic signal priority request emitters.
- Eindhoven's Hermes system is an event recorder, not connected to the radio. Location is based on sub-pavement beacons at each signalized intersection and odometer. Stop records include door opening and closing time. Records are also made of each time the bus passes a 5 km/h speed threshold, used to determine how much time is spent stopped or at crawl speed at different points on the route.
- A stop announcement system currently being installed at the CTA stores stop records on board all buses. On buses with APCs (15% of the fleet), stop records include on and off counts. At the same time, an independent 1995-vintage AVL system polls buses for their location every 40 to 70 s. Although the poll messages are recorded, they are not matched to location and are therefore unsuitable for routine operations analysis.
- At NJ Transit, an AVL vendor supplied the on-board computer with location tracking function, even though NJ Transit's system is not connected to a radio, and integrated it with APCs. Only a fraction of the fleet is instrumented; that fraction has been concentrated on one route to make operations analysis of that route possible. It features extensive and frequent event recording, including stop records, and was designed to enable later integration with the radio and other devices through a J1708 network.
- At Metro Transit, the AVL vendor supplied the on-board computer, which is connected to the radio and, on 12% of the fleet, to APCs. The radio carries both round-robin poll data (bus location when polled) for real-time monitoring and event messages. Off-line analysis will ignore the poll data except for incident investigations, and instead use event messages, including timepoint messages. Stop messages, including passenger counts, are not recorded on board, but are transmitted by radio whenever an APC-instrumented bus actually serves a stop. During periods of radio failure, event messages are recorded on board, uploaded at the end of the day, and inserted into the radio message database.
- King County Metro has both AVL and APC systems that share signpost, odometer, clock, and operator login (route/run) information, but are otherwise independent. The AVL system sends timepoint messages as well as performs round-robin polling. Service analysts at King County Metro had long relied on APC records for operations analysis; however, because of improvements in timepoint detection made around 2000, AVL timepoint data is now the preferred data source, because AVL data is implemented fleetwide, offering large and recent data samples.
- OC Transpo and Societé de Transport de Montréal (STM) have stand-alone APCs. OC Transpo is a long-time APC user and has used its APC system extensively for operations analysis as well as passenger count analysis. STM has recent in-depth experience in testing and approving new APC systems.

The fifth source of information was a 1-day workshop for vendors held on May 28, 2002. An open invitation was extended to vendors of AVL, APC, and related products, with specific invitations sent to known vendors. They were joined by panel members, representatives of several of the case study sites, and members of the project team, providing a good representation of interested transit agency staff and independent researchers. Participants other than project team members are listed in Table 3. The researchers also benefited from direct interaction with vendors referred by transit agency staff from the case study sites.

Finally, the researchers used their knowledge of the transit industry and related industries, supplemented by information received from members of the project panel.

Table 3. Vendor workshop participants.

Type	Participants
Vendor Representatives	Dirk van Dijl, ACIS Alain de Chene, Infodev Carol Yates, Orbital George Mount, NextBus Andreas Rackebrandt, INIT Neil Odle, IRIS Anil Panaghat, US Holdings Vijay Raganath, consultant with Delaware Transit Hershang Pandya, US Holdings Rohit Patel, Intelect Corporation Mike Kushner, Logic Tree
Panel Members, Liaisons, and Additional Reviewers	Jim Kemp, NJ Transit (panel chair) Wei-Bin Zhang, Univ. of California PATH program Fabian Cevallos, Broward County Transit Gerald Pachucki, Utah Transit Authority Tom Friedman, King County Metro Kimberly Slaughter, S.R. Beard Erin Mitchell, Metro Transit Yuko Nakanishi, Research and Consulting Sarah Clements, FTA Bob Casey, U.S.DOT Volpe Center Stephan Parker, TRB Eric Bruun, consultant
Case Study Site Representatives (in addition to panel members already listed)	Steve Callas, Tri-Met Kevin O'Malley, CTA Michel Thérer, STM Glenn Newman, NJ Transit

1.3.2 Analysis of Data Systems and Opportunities

The second thrust of the research was identifying actual and potential ways that archived AVL-APC data could be used and analyzing these identified ways in terms of needs for data capture, accuracy and sample size, data structures, and analysis methods. At the same time, data collection and database system designs were analyzed in terms of their capability for satisfying those needs. Based on that comparison, the researchers developed guidance regarding system design.

Another part of this thrust was the development, as a proof of concept, of one innovative data structure for analyzing service on a trunk shared by multiple patterns or lines.

1.3.3 Development and Demonstration of Improved Tools

If there is no good archived data, there appears to be no need to develop tools to analyze it. And if there are no useful tools to apply, there appears to be no need to purchase a system to gather the data. Because of this chicken-and-egg relationship between analysis tools and the practice of capturing and archiving data, the third main thrust of this project was to develop and demonstrate the use of improved analysis tools that take advantage of archived AVL-APC data.

This thrust, in its effort to accelerate the cycle of development, had two facets: TriTAPT and improved analysis tools. Delft University's software package TriTAPT contains analysis tools related to running time and passenger counting. Under the terms of this project, this software will be available with no license fee for 4 years to transit agencies in the United States and Canada. In the course of the project, it was tested by three agencies (Tri-Met, Metro Transit, and the Massachusetts Bay Transportation Authority [MBTA]); and the feedback was used to help adapt it to U.S. practice.

The second facet was the development of improved analysis tools. Improved tools for running time analysis were developed within TriTAPT, while prototype tools for analyzing passenger waiting time and crowding were developed on a spreadsheet platform.

1.4 Report Outline

This chapter describes the historical background, the research objective, the research approach, and the case study sites.

Chapters 2 and 3 review systems used to collect data. Chapter 2 covers the core vehicle location system including on-board computer and communication. Chapter 3 covers other on-board devices that can be part of a data collection system. These chapters analyze how the design of an AVL system affects the types and quality of data that it delivers.

Chapter 4 reviews actual and potential uses of archived AVL-APC data, analyzing for each use the kind of AVL-APC data it requires. With the previous two chapters, it provides the logical progression from data use to data collection system (i.e., what kind of data is needed to perform a certain type of analysis and what kind of data collection system is needed to collect that data).

Chapters 5 through 7 describe specific analysis tools that were developed in the course of this project: tools for analyzing running time, passenger waiting time, and passenger crowding.

Chapters 8 and 9 focus on passenger count data. Chapter 8 deals with schedule matching, trip parsing, and load balancing methods and includes some newly developed parsing and balancing methods. Chapter 9 deals with sampling issues and, in particular, how APC data can be used to estimate annual systemwide passenger-miles in order to satisfy NTD requirements.

Chapters 10 and 11 offer guidance on the design of automatic data collection systems (Chapter 10) and on the design of software used to store and analyze archived data (Chapter 11).

Chapter 12 discusses organizational issues associated with AVL-APC data collection and archiving. Chapter 13 offers conclusions.

Appendixes A through I, previously published as part of *TCRP Web-Only Document 23,* are case studies.

Part of this project was the development of software analysis tools. Spreadsheet tools for analyzing passenger waiting time and crowding, described in Chapters 6 and 7, are available on the project description web page for TCRP Project H-28 on the TRB website (www.trb.org). TriTAPT, which contains a suite of analysis tools including running time analysis tools developed in this project, is available with no license fee to U.S. and Canadian transit operators through the end of 2009; to request a copy including documentation, sample data files, and a set of input data conversion routines, please e-mail a request to tritapt@coe.neu.edu. The conversion routines were developed for the native format of input data used by those agencies that until now have used TriTAPT; agencies may have to adapt them to their particular input file formats.

Other products of this project were three papers published in *Transportation Research Record: Journal of the Transportation Research Board*.

- "Designing Automated Vehicle Location Systems for Archived Data Analysis" (6) contains material that is summarized in Chapters 2, 3, 4, and 10.
- "Making Automatic Passenger Counts Mainstream: Accuracy, Balancing Algorithms, and Data Structures" (7) contains material that is reproduced in Chapters 8 and 9.
- "Service Reliability and Hidden Waiting Time: Insights from AVL Data" (8) contains material that is covered in Chapter 6, as well as theoretical material that is not repeated in this report.

CHAPTER 2

Automatic Vehicle Location

TCRP Synthesis of Transit Practice 24: AVL Systems for Bus Transit provides an insightful description of AVL systems and a review of AVL's history (9). Historically, AVL was developed for two real-time applications: emergency response and computer-aided dispatch (CAD). CAD represents a major advance in complexity, because it involves matching the observed operation to the schedule. AVL has long been advertised as a means of obtaining data to be archived for off-line analysis as well. That promise, which has seen limited fulfillment, is the focus of this report.

2.1 Location Technology

In the last decade, the U.S. government's global positioning system (GPS) has become the preferred location technology. GPS receivers on vehicles determine their location by triangulation based on signals received from orbiting satellites. Location accuracy for buses is generally better than 10 m, depending on the accuracy of clocks in the GPS receivers and on whether differential corrections are used.

Because GPS requires a line of sight to the satellites, GPS signals can be lost as buses pass through canyons, including man-made canyons caused by tall buildings. Tall buildings also reflect GPS signals, causing a phenomenon called multipath that can lead to erroneous location estimation. For example, on a GPS-driven map display, buses approaching Chicago's Loop (downtown) appear to jump into Lake Michigan. In a system intended for real-time monitoring only, predictable errors such as this can be tolerated; however, for archived data intended for off-line analysis, errors like this pose a threat to data integrity. In tunnels and covered areas, GPS cannot be used unless repeaters are installed, as in NJ Transit's Newark City Subway.

Older AVL-APC systems, like King County Metro's, use a combination of beacons, which serve as fixed-point location devices, and dead reckoning for determining location between beacons, using the assumption that the bus is following a (known) route. All transit coaches have electronic odometers, making it easy to integrate odometers into a location system. Route deviations present a problem for odometer-based dead reckoning, which is one of the reasons GPS is preferred. Some AVL systems include a gyroscope, which makes it possible to track a bus off-route using dead reckoning.

Many GPS-based systems often use dead reckoning as a backup. When GPS signals indicate a change in location inconsistent with the odometer, dead reckoning takes over from the last reliable GPS measurement, until GPS and odometer measurements come back into harmony.

Odometers require calibration against known distances measured using signposts or GPS, because the relationship between axle rotations (what is actually measured) and distance covered depends on changeable factors such as tire inflation and wear.

2.2 Route and Schedule Matching

Matching a bus's trajectory to route and schedule is important for data analysis, as well as many real-time applications including CAD and real-time passenger information systems. The rate at which data is rejected for inability to match it to a route can be substantial, reaching 40% at agencies that were interviewed. Data matching was cited by many agencies as the single greatest challenge faced in making their AVL-APC data useful.

Some very simple AVL systems perform no matching; they simply display on a map where the buses are. However, most transit AVL systems include CAD, which involves real-time matching to route and schedule. Because the most demanding application in regards to matching is stop announcements (because matching errors are so apparent to the public), archived data derived from stop announcement systems should be of particularly high quality.

In traditional APC systems, which lack a real-time component such as CAD or stop announcements, data is matched off line based on signpost, odometer, and/or GPS reading recorded during operation.

Matching algorithms themselves tend to be proprietary, developed by AVL vendors or (for APC) developed in house. As a general rule, the more data available, and the greater its detail and accuracy, the better the matching accuracy is.

2.2.1 Route/Run Data and Operator Sign-In

Matching, whether in real time or off line, is more successful when the algorithm doing the matching has prior knowledge of the route and schedule the bus is supposed to be following. With the scheduled bus path known, real-time measurements are then used to verify and update the path. If door opening and closing sensors are part of the data collection system, each door opening/closing event suggests that a bus is at a stop, permitting a comparison between reported location and expected next stop location.

In newer AVL systems, the schedule and base map information used for matching are held in the on-board computer. Recognizing that schedules often change, many systems provide for schedule information to be uploaded daily. At King County Metro, where the older AVL system's on-board computers cannot hold the full schedule, buses are tracked against the schedule by the central computer; nevertheless, King County Metro devised and implemented a method for local, real-time matching. About 3 min of running time before each timepoint, the central computer radios to the bus a message indicating the odometer reading at which the coming timepoint will be located; local sensing and logic will then suffice to know when the bus has reached the next timepoint. This technique substantially improved King County Metro's success at matching timepoints.

The main source of route/run data is operator sign-in. Sign-in to radio systems is routine in the transit industry, and non-compliance is generally limited to a small percentage of operators on any given day, because operators who do not sign in can be readily detected at the control center. At King County Metro, for example, operators who fail to sign in can be called out of service for having a faulty radio, and face possible discipline if the problem turns out to be simple failure to sign in. Therefore, AVL systems connected to the radio benefit from getting relatively good-quality sign-in data.

The validity of sign-in data can also be a problem. Accuracy will be better if the system taking the sign-in accepts only valid codes for operator number, run number, and so on. Systems in which sign-in errors are not detected until off-line processing cannot benefit from operators correcting their own input errors. As an example, farebox data systems often have very high rates of sign-in error, making boarding counts and revenue difficult for agencies to attribute to route (*10*).

Data collection systems not tied to the radio, like traditional APC systems, present a challenge. Agencies are reluctant to force operators to sign in again to another system when they are already signing in to the radio, farebox, and destination sign; and, if sign-in is not necessary for any real-time operating function, compliance is sure to be an issue. King County Metro solved this problem by connecting its APCs to the radio control head, which transmits sign-in information to the (otherwise independent) APC on-board computer.

Lacking sign-in data, NJ Transit's APC/event recorder system still tries to take advantage of the scheduled runs that buses follow by using a two-step matching procedure: (1) the route/run is inferred from pull-out, pull-in, and stop records and then (2) the inferred run is used for stop-level matching. If matching fails, the system may guess a different route/run.

Metro Transit eases the burden on operators while improving accuracy by providing automated sign-in, communicated by wireless link during pull-out, based on vehicle-block assignments made overnight. In its new AVL-APC system, operators are asked only to verify and correct their sign-in information.

Houston Metro improves the accuracy of its route/run data by comparing sign-in data with payroll data as part of routine post-processing. A semi-automated procedure allows an analyst to make corrections if there seems to be a simple, correctable error, such as a miskeyed run code.

2.2.2 Base Map Accuracy

Without data matching as a driving application, transit agencies have little need historically for an accurate stop database. Many agencies have no stop database, because they do not own the stops (i.e., the sidewalk space and signs) and because routes and schedules are detailed only to the timepoint level. Before AVL, stop databases only had to be accurate enough for operators and maintenance personnel to locate the stop. However, automatic applications do not forgive errors and omissions the way manual data collection can. Generally, agencies implementing AVL have needed to make a major effort to correct their stop location database. Some agencies and vendors have used dedicated crews to field map all stop locations using mobile GPS units. Buses themselves can be configured to be those mobile units.

The 2002 case study of NJ Transit emphasizes the importance of having a good base. On patterns for which at least 90% of the reference locations are coded to within 300 feet of actual, NJ Transit's matching algorithm was able to match 81% of the trips to a scheduled trip and pattern, in contrast to a 65% matching rate overall. Starting with a well-calibrated GIS base map based on aerial photography can reduce the burden of field mapping. Equally important is maintaining the stop location file for both temporary and permanent changes. Some large agencies report changing 5% of their 10,000 stops each year.

While AVL-APC systems need accurate bus stop locations, the location data that AVL-APC systems supply can also be used to improve the base map. One APC vendor recommends comparing average GPS coordinates at observations of a stop to its reference coordinates, updating the base map if the observations are consistent.

One complicating factor is that different enterprise systems—scheduling, facilities, transportation, passenger information systems such as on-line trip planners, and traffic management—use the term "bus stop" for different functions, leading to slightly varying definitions that often entail differing locations. A transit agency can have four or more definitions of stop location (*11*):

- Intersection or landmark (e.g., Third and Main)
- Intersection quadrant
- Nominal coordinates along an ideal route, often following the roadway centerline
- Coordinates of the point on the curb closest to the bus stop sign

Additionally, the complication of determining coordinates, which may differ between location system and the base map, introduces a possible calibration problem. A recently published U.S.DOT report on location referencing offers valuable guidance for making location accurate and consistent across data systems (*12*).

2.2.3 Schedule Integration

Thanks to the near-universal adoption of automated scheduling, route and schedule data is always imported from the scheduling system. The schedule database tends to be accurate and carefully maintained because of its critical role in operations and payroll.

In large cities, transit schedules tend to be extremely dynamic, with route, vehicle, and operator schedules changing almost daily. It is therefore vital to the performance of AVL to have a mechanism for keeping the schedule up to date. Many AVL systems reload the entire schedule to bus on-board computers daily at pull-out.

Several transit agencies have reported difficulties in integrating the schedule database with AVL, sometimes delaying a project for years. The desire for a standard interface was echoed by many transit agency and vendor representatives. Even though there are only two major schedule software suppliers, databases tend to be highly customized to each transit agency's particular routing, schedule, and work rules practices; therefore, a standard interface, even for a single scheduling system vendor, can be elusive.

In one case, the problem that had to be overcome was that the AVL vendor's software expected that the schedule database would have details concerning the path taken by buses during pull-ins, pull-outs, and deadheads, but the scheduling system vendor's database excluded those details. Another problem is that schedule databases define routes as a series of timepoints, not stops, while an AVL or APC system doing stop-level tracking sees a route as a sequence of stops.

2.2.4 Control-Ordered Schedule Changes

Route and schedule matching becomes complicated when buses do not exactly follow their assigned block. Examples are when a bus breaks down, when buses are short-turned or inserted into the schedule to try to balance headways, or when buses swap duties. In such cases, capturing information about schedule changes ordered by dispatchers can ease the task of matching.

If the change is simply that a bus assumes a new block or run, that information can be captured through (a fresh) sign-in. Otherwise, to the researchers' knowledge, no method has yet been developed for capturing control-ordered schedule changes in a form suitable for automated processing. At King County Metro, for example, the AVL/radio database includes controller logs indicating changes to bus and operator assignments; however, those records need to be interpreted manually.

2.2.5 End-of-Line Identification

End-of-line operations can be both complex and unpredictable, making a trip's start and end times difficult to identify. One reason such identification is difficult is that terminals are often located where GPS accuracy is worst—near tall downtown buildings or in a covered terminal.

A second reason is the unpredictability of operations at route ends. Operators approaching the end of the line with an empty bus may feel free to deviate from the prescribed route (e.g., to stop at a sandwich shop, thereby spending their layover at a different location). At the terminal, an operator getting in and out to adjust a mirror can be mistaken as passengers getting off and on and a stop being served. Operators may open and close doors several times to let passengers board during layover periods, which makes door closing an inadequate criterion for inferring departure from the stop. A vehicle jockeying for position in a layover area may be mistaken for an early departure.

For these reasons, several agencies report treating first and last segment running time data with some skepticism. Some agencies simply exclude first and last segments from running time analysis, forcing them to assume a fixed running time on those extreme segments. Not being able to track operations from the start to the end of a line compromises the integrity of route-level running time analyses such as determining periods of homogeneous running time and the sufficiency of recovery time.

Agencies have used various means to improve end-of-line identification. King County Metro made its tracking algorithm ignore bus movements in layover areas that occur more than 3 min before scheduled departure time, because bus movements in staging areas were being falsely detected as trip starts. Staff at several agencies recommend locating signposts (real or virtual) not at route terminals, but a few minutes' travel from the terminal. This arrangement resolves some of the matching issues but leaves running time and schedule adherence at the route endpoints unmeasured. (Stop-level data collection effectively accomplishes the same thing.) Records made during the layover, and algorithms using those clues can help reduce end-of-line matching errors.

End-of-line complications also affect passenger count analysis, as discussed in Chapter 8.

2.3 Data Recording: On- or Off-Vehicle

One essential distinction within automatic data collection systems is whether data is recorded in an on-board computer or in an off-vehicle central computer to which messages are sent via radio. Historically, AVL systems have been tied to the radio system and have used that connection for off-vehicle data recording. Stand-alone APC systems, in contrast, record data on board. From the viewpoint of a data archive, radio-based systems are limited by radio channel capacity, while on-board storage imposes no meaningful limit on either the number or detail of data records.

Until the mid-1990s, on-board data storage was expensive and was therefore avoided. Since then, the cost of adding storage capacity to on-board computers has ceased to be a significant factor in system design. Therefore, newer systems, which sometimes blur the traditional AVL-APC distinction, can be designed with either on- or off-vehicle data recording, or both. In the mid-1990s, Tri-Met made the significant step of specifying that its new AVL system both transmit messages by radio to serve real-time applications and store event records on board for off-line analyses. Because of radio channel limitations, radio messages are sent only for exceptions (a bus is more than 3 min late or off route, or an event of interest occurs); on-board recording, in contrast, has no such limitation.

2.3.1 Radio Messages and Record Types

Radio systems use a wide area network (WAN) to manage communication between a central computer and on-board computers and radios. They are licensed by the Federal Communications Commission (FCC) and have a limited number of radio channels. Limitations are most severe in large cities, where the demand for radio channels (for police, fire, taxi fleets, etc.) is greatest; unfortunately, large cities also have the larger bus fleets and therefore need more radio channels. Transit agencies typically dedicate some channels to voice and others to data; this research concerns the data channels. Because there may be hundreds of buses per radio channel, the radio traffic has to be managed to fit within the available capacity.

Polling Records

Most real-time AVL systems use round-robin polling to track their vehicles. The polling interval depends on the number of vehicles being tracked per radio channel; 40 to 120 s is typical. Within each polling cycle, every vehicle is polled in turn, and the vehicle responds with a message in a standard format. Round-robin polling is an effective protocol for avoiding message collisions; however, the need to transmit messages in both directions, with a time lag at either end for processing and responding, means that a significant amount of time—on the order of 0.5 s—is needed to poll each bus. The polling cycle is therefore limited by the number of buses being monitored per radio channel.

A polling message includes ID codes (for the vehicle, its run or block, and perhaps its route) and various fields for location data. Location fields depend on the location system used. For a beacon-based system, they include ID of the most recently passed beacon and odometer reading. For GPS systems, GPS coordinates will be sent, perhaps with odometer reading as well. The polling message often includes other fields such as an operator-activated silent alarm and mechanical alarms such as "engine overheated."

Polling provides location-at-time data (i.e., the location of the bus at the arbitrary time at which it is polled). However, the much more useful time-at-location data (i.e., time at which a bus passes a point of interest such as a stop or timepoint) is the format needed to analyze schedule adherence and running time. While polling data can be interpolated to get estimates of time-at-location data, such interpolation can involve significant approximation error, especially when buses are traveling at low speeds because of traffic or stopping. In principle, if the polling interval were very short, approximations would become insignificant; however, radio channel capacity limitations make short polling intervals impractical. (Many systems can switch to a short polling cycle for particular buses in an emergency, but that can only be done to the detriment of other buses' polling interval.)

The researchers found no examples of transit agencies extracting time-at-location information from polling data or basing any analysis of running time or schedule adherence on it. The only off-line use found for polling data was for detailed investigations of incidents using playback.

Event Records

In addition to round-robin polling, WANs also support messages initiated at the vehicle, generically called "event messages." Each event record has a code and specified format. Modern AVL systems can have 100 or more different types of event records.

Messages initiated by on-board computers are likely to collide—that is, one bus will try to send a message while the channel is busy with another message. WANs manage this kind of network traffic problem in various ways, such as by having messages automatically re-sent until a receipt message is received. This need to manage traffic limits the practical capacity of radio-based communication, because, with randomly arising messages, the channel has to be unoccupied a relatively high fraction of the time (unlike with round-robin polling) to provide an acceptable level of service. In the face of limited channel capacity, then, radio-based systems have to be designed in a way that limits the frequency and length of messages sent.

Timepoint Records

In most AVL systems, the timepoint event, indicating a bus's arrival or departure from a timepoint, is the most frequent event record used for archived data analysis. The event can be defined in various ways, depending on the location system. Where GPS is used and door switches are not, it is common to report when the bus first reaches a circular zone (typically a 10-m radius) around the stop. The timepoint record may also include the time the bus left that zone. In principle, timepoint records could also include fields indicating when doors first opened and last closed; however, the researchers are not aware of any radio-based systems incorporating door information.

The level of detail of timepoint records affects their accuracy and value for off-line analysis. For example, some running time and schedule adherence measures are defined in terms of departure times, others in terms of arrival times, and others involve a difference between arrival time at one point and departure time at a previous timepoint. Off-line analysis therefore benefits from having both arrival and departure times recorded, particularly if operators hold at timepoints. Records of when buses enter and depart a stop zone are only approximations of when buses arrive and depart the stop itself. Errors can be significant in congested areas where traffic blocks buses from reaching or pulling out of a stop. Detail on door opening and closing, and on when the wheels stop and start rolling, can help resolve ambiguities and make arrival and departure time determination more accurate.

Stop Records

Stop events are much more frequent than timepoint events, and therefore, far more demanding of radio channel capacity if transmitted over the air. Therefore, most AVL-APC systems collecting data at the stop-level store stop records in the on-board computer, uploading them overnight. However, some systems, including Metro Transit's, find enough radio channel capacity to send stop messages over the air, though only for the subset of the fleet (under 20%) instrumented with APCs.

The data items typically included in a stop record—in addition to the usual time stamp, location stamp, vehicle IDs, and door switches—are door opening and closing times and (if available) on and off counts. If routes are tracked by the on-board computer, as is the case with stop announcement systems, the stop record will include stop ID in addition to generic location information; otherwise, the data is matched in later processing.

Other Event Records

Timepoint and stop events are the frequent events that most off-line analysis relies on. Other event records can be valuable, either in their own right or because of the clues they offer for matching.

Events whose records can be helpful for matching include the operator signing in, bus passing a beacon, bus going off route, and engine being turned off or on. "Idle" event records, indicating that a bus has not moved for a certain amount of time, are useful for confirming bus behavior at layover points or timepoints. Heartbeat records, written every 30 to 120 min, help confirm that the bus's data recording system is working.

Events of direct interest for off-line analysis include wheelchair boardings or alightings, bicycle mounting, various delay types (e.g., drawbridge, railroad crossing), and pass-ups. Most such event messages are manually triggered by the operator; although some (e.g., wheelchair lift use) can be automatically generated.

2.3.2 On-Board Data Recording and More Message Types

In contrast to radio-based data recording, on-board data recording essentially offers unlimited capacity (because on-board data storage is so inexpensive it is easily obtainable). It is also more robust, not being subject to radio system failure. (Some radio-based systems, such as Metro Transit's, include backup on-board data storage during periods of radio failure to prevent data loss. Event records stored on board are uploaded at pull-in using the radio system and merged into the event database.)

Radio Integration and Operator-Initiated Data

Expecting operators to key in event data that is used for data archiving only is considered unrealistic. Therefore, event records are generally limited to what can be automatically generated, unless the on-board computer is connected to the radio, in which case it can also capture operator sign-in and operator-initiated events. Several AVL-APC systems feature this arrangement, which helps facilitate matching and yields a richer database. On-board integration with the radio control head is recommended even if there is no real-time AVL.

Off-line, radio-based event records also can conceivably be integrated with records uploaded from the on-board computer. However, the researchers are not aware of this design being used.

Interstop Records and Detailed Tracking

With on-board data collection, data can also be collected between stops. One data collection mode is to make records very frequently (e.g., every 2 s); this mode uses buses as GPS probes, which can enable such special investigations as studying the bus's path through a new shopping center or studying bus movements in a terminal area. Frequent interstop records also offer information on speed and acceleration.

An alternative data collection mode is to write event records for defined events that can occur between stops, such as crossing a speed threshold. To measure delay, Eindhoven's system records whenever a bus's speed rises above 5 km/h or falls below 4 km/h. (Using differing thresholds prevents oscillation when the bus is traveling at the threshold speed.) Records for a variety of speed thresholds could be useful for analyzing speed profiles. Tri-Met's system tries to capture maximum speed between stops, both as a measure of traffic flow and as a safety indicator. It tracks speed continuously, storing the greatest speed since the last stop in a temporary register; then at each stop, maximum speed since the previous stop is recorded.

Data Uploading

When data is recorded on vehicle, there has to be a system for uploading the data from the on-board computer to the central computer. Newer systems usually include an automatic high-speed communication device through which data is uploaded daily when buses are fueled. Older systems such as Tri-Met's rely on manual intervention, such as exchanging data cards or attaching an upload device, which adds a logistical complication.

The absence of an effective upload mechanism can render an otherwise promising data collection system useless for off-line data analysis. One transit agency has a new stop announcement system that records departures from every stop. However, the data is overwritten every day, because the data logging feature was intended for system debugging, not data archiving. To make it an archived data source, the agency would have to either exchange cassettes nightly, something it deemed impractical, or invest in a high-speed data transfer link, something it found too expensive.

Communication is also needed from the central computer to the on-board computer, whether for occasional software upgrades or daily schedule updates. In general, the communication method used for upload is used for download as well.

2.3.3 Exception-Only Data Recording

Some AVL systems send timepoint messages only if a bus is off schedule, partly to limit radio traffic, and partly because controllers are usually interested only in service that is off schedule. For example, Tri-Met's AVL-APC system transmits only exception messages by radio, while saving a full set of records on board for later analysis.

However, if exception data is all that is available for off-line analysis, analysis possibilities become severely limited. For example, if only off-schedule buses create timepoint records, running times can only be measured for off-schedule buses. Researchers at Morgan State University tested the feasibility of data analysis using exception data from bus routes in Baltimore, MD (*13*). Because they had records only on buses that were outside an on-time window, they focused on next-segment running time for buses that arrived at a timepoint early or late. If the bus reached the next timepoint "on time," the researchers had to guess when within the on-time window the bus arrived. Also, because the system archived only exception messages, it was impossible to know whether a missing record meant that a bus was on time or that the radio system had failed. In an interview, the Morgan State researchers stated that while the Mass Transit Administration (MTA) had asked them to systematize the way they transformed the raw data into records that would support analyses of running time and schedule adherence, they felt it was impossible given the frequent need for assumptions to make up for missing data.

Fortunately, advances in radio technology have reduced the pressure to limit data collection to exceptions. As an example, CTA's real-time AVL system, specified in 1993 to provide location data only if buses were off schedule, was modified in 2002 so that buses transmit location data regardless of whether they are off schedule.

2.4 Data Recovery and Sample Size

Automatic data collection systems do not offer 100% data recovery. Traditional APCs have the worst record; a 1998 survey found net recovery rates for APCs ranged from 25% to 75%, with newer systems having better recovery rates (*14*). In 1993, the Central Ohio Transit Authority (COTA) reported that, with 11.9% of the fleet APC equipped, it netted on average five usable samples per assignment each quarter (*15*). That represents a 6.1% sampling rate, for a net recovery rate of about 50%. Only a small part of that loss was due to mechanical failures, as mechanical reliability was reported to be well above 90%.

Less is known about data recovery rates for radio-based AVL systems. King County Metro recovers AVL data from about 80% of its scheduled trips. However, with the entire fleet instrumented, data recovery rates are not so important with AVL unless there is systematic data loss in particular regions.

Inability to match data in space and time is the single most cited reason for rejecting data. There are other reasons, including malfunctioning on-board equipment, data being out of range, radio failure, and so forth. Imbalance in passenger counts, another common problem that forces data to be rejected, is covered in Chapter 8.

The effective sampling rate of an AVL-APC system is the fleet penetration rate multiplied by the data recovery rate. If 10% of the fleet is equipped, and data is recovered from 70% of the instrumented vehicles, scheduled trips will be observed, on average, 10% * 70% = 7% of the number of times they are operated. In a 3-month period containing 65 weekdays, 13 Saturdays, and 13 Sundays, average observations per scheduled trip will be 4.5 for weekday and just under 1 for Saturday and Sunday trips.

An average sampling rate can mask significant variations across the system. When fleet penetration is small, logistical difficulties coupled with the vagaries of data recovery failure often result in some scheduled trips being sampled well more than the average number of times, and others perhaps going completely unobserved. Management of the rotation of the instrumented vehicles, then, becomes another important factor in determining whether needed data will be available.

In an example drawn from the project's case studies, an audit of King County Metro's Fall 1998 sign-up (a 4-month period) found on average six valid APC observations per weekday scheduled trip, for a sampling rate of 7.5%. With about 15% of the fleet instrumented, that represents a data recovery rate of 50%. Coverage across the schedule was variable. King County Metro recovered at least one valid observation on 97% of their scheduled weekday trips, and at least three valid observations for 83% of its scheduled weekday trips. On the weekend, 85% of scheduled trips had at least one valid observation. In 2002, King County Metro reported that its recovery rate had improved to the 60% to 70% range.

In general, a small sample size is sufficient to reliably estimate the mean of a quantity with low variation such as running time on a segment or demand on a particular scheduled trip. In contrast, a large sample size is needed to examine variability or extreme values, such as the 90-percentile running time or load. A very large sample size is needed to accurately estimate proportions, such as proportion of departures that are on time (*16*).

Analyses that aggregate scheduled trips into periods, or that aggregate routes, have the advantage of a larger sample size than analyses of individual scheduled trips. For this reason, many customary analysis methods, developed in a limited-data environment, find results for the period rather than the trip, and for the system rather than the route.

A large sampling rate allows more timely analysis of data. Over a long enough period of time, even a small sampling rate will yield a large number of observations. However, for management to react promptly to demand and performance changes, or measure the impact of operation changes, analysts need recent data. Therefore, tools used in the active management of a dynamic system will benefit from the high sampling rate that follows when the entire fleet is equipped.

Because leader-follower or headway analysis requires valid observations of consecutive pairs of trips, the number of valid headways one can expect to recover is proportional to the square of the data recovery rate and the correct assignment rate. For example, if the data recovery rate is 70% and if a request to instrument a given route results in 90% of the trips on that route having an instrumented bus, one can expect to observe only $(0.7)^2 * (0.9)^2 = 40\%$ of the day's headways on that route. If there is a realistic possibility of trips overtaking each other, having anything less than data from all the operated trips casts some doubt on calculated headways.

CHAPTER 3

Integrating Other Devices

While automatic vehicle location, automatic passenger counting, and farebox systems began as the single-purpose systems their names imply, their value as data collection systems soon became apparent. Using the expanded definition of AVL as an automatic data collection system that includes location measurement, this section describes other devices that can be integrated into the data collection system.

3.1 Automatic Passenger Counters

Unlike AVL, APCs have always been designed with archived data in mind. Valuable reviews of the history of APCs are found in reports by Levy and Lawrence (17), Boyle (14), and Friedman (18). APCs use a variety of technologies for counting passengers, including pressure-sensitive mats, horizontal beams, and overhead infrared sensing. Automatic passenger counting has not yet seen widespread adoption primarily because of its cost and the maintenance burden it adds. Where adopted, APCs are typically installed on 10% to 15% of the fleet. Equipped buses are rotated around the system to provide data on every route. However, technological advances may soon make APCs far more common.

The term "APC" can refer to a full data collection system or to simply the passenger counter as a device within a larger data collection system. Historically, APCs were implemented as full, independent systems that included location measurement and stop matching. In spite of the emphasis their name gives to passenger use data, they not only counted passengers but also provided valuable operation data that supported analysis of running time and schedule adherence; in effect, they doubled as (non-real-time) AVL systems. Canadian transit agencies have been particularly active in exploiting APC data. OC Transpo, the Toronto Transit Commission, Winnipeg Transit, Tri-Met, and King County Metro are among the agencies that have long benefited from routine reports on passenger loads, running time distribution, and on-time performance from APC systems.

Since the mid-1990s, the trend has been for APCs to become simply a component in a larger data collection system that includes automatic vehicle location. Location and stop matching is the duty of the vehicle location system, which operates out of a main on-board computer called the "vehicle logic unit." The passenger counter includes sensors and a dedicated on-board computer called an APC analyzer that converts sensor information into passenger counts. Each time the bus leaves a stop, the APC analyzer closes out a record and transmits its on-off counts to the vehicle logic unit. From there, the data is treated like data from any other device in the data collection system—either stored on board for later upload, or transmitted by radio in a stop event message.

By integrating APCs into an AVL system, the marginal cost of passenger counting drops dramatically. Tri-Met, already committed to fleetwide AVL and using one of the simpler APC designs, finds the marginal cost of adding passenger counting to be in the range of $1,000 to $3,000, versus unit costs of $5,000 to $10,000 often cited for stand-alone systems. This marginal cost is low enough that Tri-Met includes APCs in all new coach purchases. With 65% of its fleet already equipped, it is the only large transit system with APC penetration beyond a small sample of the fleet. Benefits of a large APC sample are discussed in Section 10.4.

APC counting accuracy depends on the technology used, the care used in mounting and maintaining sensors, and algorithms used to convert sensor data into counts. The accuracy of finished counts also depends on the effectiveness of stop matching and identifying the end of the line, subjects discussed earlier; it also depends on algorithms used for screening, parsing, and balancing, which are covered in Chapter 8.

3.2 Odometer (Transmission Sensors)

As mentioned earlier, all buses have electronic transmission sensors that serve as odometers, giving a pulse for every axle rotation. Integrating the transmission into an automatic data collection system provides the basis for dead reckoning and enables data on speed to be captured. Estimating speed from GPS measurements is not reliable (except over substantial distances) because of random measurement errors. Therefore, odometer input is preferred to determine when a bus's wheels start rolling after serving a stop.

Some AVL systems also integrate a gyroscope, which indicates changes in heading. Gyroscope readings will support off-route dead reckoning and can aid in matching because they detect turns.

3.3 Door Switch

APC systems in North America always include door switches to help determine when a bus is making a stop. Even when a system does not include passenger counting, door switches can be a valuable means of location matching. If a system repeatedly shows doors opening at a location not coded in the base map as a stop, that information can be used to update the base map. In some Dutch transit systems, door sensors are used to distinguish time spent holding (bus is resting at a stop, ahead of schedule, doors closed) from dwell time spent serving passenger movements.

APC vendors in Latin America know that buses there often operate with doors open, rendering door switch readings useless.

3.4 Fare Collection Devices

The traditional electronic farebox has limited data storage capacity, creating one record per one-way trip with simply a count of the trip's boardings and revenue. Farebox manufacturers have historically been reluctant to allow their machines to communicate with other on-board devices, citing the need to prevent fraud by keeping revenue-related information secure. Limited integration schemes, such as sharing a common operator log-in and interface to the destination sign (to indicate change in route/direction), have been applied at a few transit agencies. Recent products advertise J1708 compatibility (see Section 3.6).

A new development is the transactional farebox, which produces a time-stamped record for each transaction. If the fareboxes are networked to a smart bus system, transaction data can be transmitted along the data bus and collected as part of the AVL event stream.

If fareboxes are not integrated, off-line integration of the farebox's own data stream with archived AVL data, in principle, should be possible, with matching performed on the basis of vehicle ID and time. A research project for the CTA attempting to prove this concept found several obstacles to integrating the data sources. Farebox and AVL clocks were not synchronized. Also, the fare transaction data contained many unexplained anomalies. For example, one would expect that people transferring from one route to another (the farebox transaction data includes route transferred from) would board the second bus where the two routes intersect, providing a means of verifying a stop matching; however, the data shows such transfers occurring at multiple stops, some a good distance from the transfer point. Because farebox data is not usually analyzed at this level of detail, its quality in many respects has not mattered before. Improving the quality of farebox data will be a challenge to efforts to integrate it with AVL-APC data. Once this challenge is met, however, it offers the prospect of a rich data source from 100% of the fleet.

While fareboxes cannot directly measure load because they do not register passengers both boarding and alighting, there are methods of estimating load based on the historical symmetry between the boardings pattern in one direction and the alightings pattern in the opposite direction (*19*). As contactless smart cards penetrate the market, card readers, some day, may be able to count passengers alighting as well as boarding.

Transaction data in which the fare medium is electronic offers the possibility of tracking linked trips and analyzing transfers, by linking records with a common user ID. Knowing the pattern of where and when a particular farecard was used to enter the system allows estimation of the cardholder's trip pattern to be made, based on round trip symmetry. The viability of this approach has been demonstrated in the New York subway system (*20*) and in the multimodal Helsinki transit system (*21*). A research project is currently under way, with promising results, using Dublin bus data in which farecard transactions are all station stamped.

In the near future, Metro Transit plans to introduce smart cards, bypassing the farebox, with smart card readers integrated into its vehicle location system. This arrangement may finally produce the long hoped-for benefits of integrating fare collection with vehicle location.

3.5 Other Devices

3.5.1 Radio Control Head

As discussed in Section 2.2.1, integrating the control head ensures that the AVL data stream gets both sign-in data and data messages sent by the bus operator to the control center.

3.5.2 Passenger Information Systems

Passenger information devices, including the destination sign and next stop announcements, can be integrated with a location system; however, because they become consumers, not suppliers, of information, they add nothing directly to a data archive. Their integration does bring an indirect benefit to archived data by increasing the pressure for the system to match routes and stops accurately.

3.5.3 Wheelchair Lift

Lift sensors can initiate a location-stamped record of lift use, a valuable piece of information for off-line analysis of both ridership and running time. In the survey, no examples of lift sensors being used in AVL systems were found; lift data, where available, came from operator-initiated messages sent by radio.

3.5.4 Silent Alarm

AVL systems usually include an operator-initiated silent alarm for emergencies. Its recording is valuable for incident investigations.

3.5.5 Mechanical Sensors

Transit buses have had electronic controls and monitoring systems for many years (22). Besides transmission sensors, such systems include, for example, engine heat and oil pressure sensors. AVL systems have sometimes taken input from such sensors, triggering mechanical alarms for high temperature or low pressure; however, false positives have been too frequent and therefore no use has been found for mechanical alarms.

It has been suggested that mechanical data incorporated into a data archive (as opposed to triggering real-time alarms) may yield valuable insight into maintenance needs.

3.6 Integration and Standards

On-board devices can be integrated in an ad hoc fashion, or following a systematic integration standard. Standardization helps promote the cost and efficiency benefits of modularity and re-use.

3.6.1 SAE Standards and Smart Bus Design

Systematic on-vehicle integration of devices has been called the "smart bus" concept. The principal on-vehicle device is the main data collection computer or vehicle logic unit. In principle, the vehicle logic unit and other devices are each connected to a twisted pair of copper wire running the length of the vehicle and called the "data bus." (In practice, there are often a few devices, such as GPS receivers, connected directly to the vehicle logic unit instead of via the data bus. Wireless networking has also become possible as a substitute for a physical data bus.) The network consisting of the data bus and attached devices is a local area network (LAN, often called a "vehicle area network" (VAN). Devices broadcast messages on the network when triggered by an event; other devices receive or ignore the message, depending on how they are programmed. A communication protocol governs message types and manages traffic on the VAN. Nearly all the relevant devices manufactured today comply with the J1708 family of standards (23, 24) published by SAE, the integration standard used in many AVL systems. Some AVL integrators use a proprietary VAN protocol that they claim handles message traffic better.

An advantage of integration is providing operators with a single interface and a single sign-in, which can be shared (in principle) with such disparate devices as the radio, the event recorder (a function usually taken by the vehicle logic unit), the farebox, the stop announcer, and the destination sign. In practice, fareboxes are rarely integrated and therefore have a separate interface.

The most significant integration is the integration of APC with AVL, first implemented in the mid-1990s at Tri-Met. The AVL vendor provided a smart bus with location tracking; therefore, the APC subcontractor had to provide only the passenger sensors and APC analyzer, relying on the AVL system for stop matching. Another example is a stop announcement system vendor who provides the smart bus system, with APCs added as a supplemental device.

While it may seem obvious that AVL, APC, and stop announcement systems should share location systems and on-board computers, their integration is only a recent development. In fact, for various cost and contract reasons, separate location systems are still being installed independently at some transit agencies that lack the smart bus foundation, with buses having multiple GPS receivers, multiple on-board computers, and, in spite of efforts to avoid it, multiple operator interfaces. Using an open, standards-based smart bus design when procuring an AVL, APC, stop announcement, or event recording system substantially lowers the marginal cost of adding the other functions and provides flexibility for later procurements.

3.6.2 Other Integration and Standards Efforts

Transit Communications Interface Profiles (TCIP)

The TCIP project, started in 1996, is a standards development effort sponsored by the U.S.DOT's Joint Program Office for Intelligent Transportation Systems (ITS). Its mission has been to define the data elements and message sets that can be specified as an open data interface for transit data interchange activities. Phase 1, completed in 1999, established a transit ITS data interface "Framework" and eight "Business Area Object Standards." Phase 2, completed in June 2001, built on the work of Phase 1 by developing the transaction sets, application profiles, and guidebooks required to test and implement TCIP.

Some of the pertinent TCIP developments include

- The definition of automatic vehicle location objects, including compass bearing parameter, current time, current date,

trip distance, position, velocity vector, total vehicle distance, and milepost identification;
- The definition of conformance groups that consist of a list of objects required to support a specific function. Conformance groups were defined for dead reckoning, triangulation, GPS, etc.;
- The development of a standard on spatial representation objects, including a data dictionary and the definition of message objects set; and
- The development of a standard for on-board objects, also including a data dictionary and the definition of message objects set.

TCIP object definitions include most schedule features, which are important for matching a vehicle and data collected on it to the route and trip on which it is operating. Some AVL and APC vendors have adopted the TCIP standard for their schedule data. However, TCIP has not yet been widely adopted by transit agencies as part of their product specifications and Requests for Proposals. Moreover, there are aspects of schedules that are not covered in TCIP, such as when a bus is simultaneously discharging passengers from an inbound trip and picking up passengers for an outbound trip. Among the people working to improve the TCIP standard are those involved with AVL and APC data.

Location Referencing Guidebook

The recently published *Best Practices for Using Geographic Data in Transit: A Location Referencing Guidebook* (*12*), funded jointly by FTA and the Transit Standards Consortium, promotes effective practices in the exchange and use of spatial data, including stop and route definitions.

FTA National Transit GIS Initiative

The FTA initiated in the mid-1990s a National Transit Geographic Information System to develop an inventory of public transit assets in the United States. Although the effort focused primarily at a high national level, it was also designed to encourage more use of GIS tools by transit systems. Its 1996 report discusses the design of bus stop and route databases as well as specifications for data exchange involving GIS databases (*25*). This effort was to some extent a precursor to the *Location Referencing Guidebook* project discussed earlier.

CHAPTER 4

Uses of AVL-APC Data

This chapter includes traditional analysis methods as well as new tools developed by agencies using archived AVL-APC data. In addition, part of this project's survey/interview process included asking for suggestions of how AVL-APC data might be used, and the researchers also developed some concepts.

To a large extent, performance analysis can be driven by a performance-monitoring protocol that specifies measures or indicators that are to be reported. Transit agencies have experimented with a large number of performance measures. AVL or APC data apply to many of the performance measures listed in *TCRP Report 88: A Guidebook for Developing a Transit Performance-Measurement System* (*26*). *TCRP Report 100: Transit Capacity and Quality of Service Manual* (*27*) also suggests several AVL- and APC-related measures for use as service quality indicators.

A list of current and potential uses of AVL-APC data is given in Table 4. They are discussed in detail beginning in Section 4.3, following general discussions of data uses in an era of automatic data collection (Section 4.1) and of data needs for different analyses (Section 4.2). Section 4.3 includes several examples of analysis reports; many more can be found in the accompanying case studies. Both TCRP (*14*) and the Canadian Urban Transit Association (*17*) have published syntheses of practice that include numerous samples of reports generated from archived AVL and APC data.

4.1 Becoming Data Rich: A Revolution in Management Tools

The transit industry is in the midst of a revolution from being data poor to data rich. Traditional analysis and decision support tools required little data, not because data has little value, but because traditional management methods had to accommodate a scarcity of data. Automatic data-gathering systems do more than meet traditional data needs; they open the door for new analysis methods that can be used to improve monitoring, planning, performance, and management.

Transit agencies—such as Tri-Met, King County Metro, OC Transpo, and HTM—that have good, automatically collected operational data are finding more and more uses for it. At first, agencies may look to an automatic data collection system only to provide the data needed for traditional analyses. But, once they have the larger and richer data stream that AVL and APCs offer, they think of new ways to analyze it, and they want more. Eventually, their whole mode of operation changes as they become data driven. One APC vendor explains, "We're selling an addiction to data."

Five trends in data use have emerged from the paradigm shift from data poor to data rich:

- Focus on extreme values
- Customer-oriented service standards and scheduling
- Planning for operational control
- Solutions to roadway congestion
- Discovery of hidden trends

4.1.1 Focus on Extreme Values

Traditional methods of scheduling and customer service monitoring generally use mean values of measured quantities, because mean values can be estimated using small samples. However, many management and planning functions are oriented around extreme values and are, therefore, better served by direct analysis of extreme values such as 90th- or 95th-percentile values. These extreme values now can be estimated reliably because of the large sample sizes afforded by automatic data collection. Three examples follow:

- **Recovery time** is put into the schedule to limit the probability that a bus finishes one trip so late that its next trip starts late. Therefore, logically, scheduled half-cycle times

Table 4. Decision support tools and analyses and their data needs.

Function	Tool/Analysis	Record Type Needed	Data Detail or Analysis Capacity Needed	Sampling Rate Needed
Targeted Investigations (complaints, disputes, incidents)	General view	Any	Incident codes, control messages.	100%
	Trip recorder view (speed, acceleration)	Interstop	Maximum speed; records every 2 s or more to measure acceleration/deceleration rates.	100%
Analyzing and Scheduling Running Time	Route and long segment running time and scheduling	Timepoint	Endpoint data needed for route running time. Both arrival and departure times are helpful.	10% for mean running time; 100% for analysis and scheduling based on extreme values
	Stop-level running time and scheduling	Stop	Door open and close times, incident codes, control messages, on and off counts.	
	Running time net of holding time	Stop		
	Speed and traffic delay	Stop, interstop		10%
	Dwell time analysis	Stop	Door open and close times, on and off counts, farebox transactions, incident codes.	10%
Schedule Adherence and Long-Headway Waiting	Schedule deviation	Timepoint	Both arrival and departure times are helpful.	100%
	Waiting time (long headway)	Stop		
	Connection protection	Timepoint	Arrival and departure times, control messages, farebox transactions.	
Headway Analysis and Short-Headway Waiting	Plotting successive trajectories	Timepoint		100%
	Route-level headway analysis	Timepoint		
	Waiting time (short headway)	Stop		
	Bunching (load/headway) analysis	Stop	Current on and off counts.	
Route Demand Analysis	Demand along a route	Stop	On and off data.	10% for mean values; 100% for analysis and scheduling based on extreme crowding
	Demand over the day and headway/departure time determination	Stop	On and off data.	
	Passenger crowding	Stop	On and off data.	Near 100%
	Pass-ups, special uses, and fare categories	Stop	Incident codes, farebox transactions.	100%

Mapping	Geocode stops, verify base map	Stop	10%
	Map bus path through shopping centers, new subdivisions, etc.	Interstop	10%
Miscellaneous Operations Analysis	Acceleration and ride smoothness	Interstop	100%
	Mechanical demand	Stop, interstop	100%
	Terminal movement	Interstop	100%
	Control messages	Varies	Incident codes and control messages.	100%
	Operator performance	Varies		100%
Higher Level Analysis	Before–after study	Varies	As required by the type of analysis.	varies
	Special event/weather analysis			
	Trends analysis			
	Aggregation and comparison over routes, including systemwide passenger-miles	Varies	Multiroute analysis capacity	
	Transfer and linked-trip analysis	Stop	Farebox transactions with card IDs.	100%
	Shared route analysis, including headways and load on a trunk	Varies	Data structures for shared routes.	
	Geographic demand and service quality analysis	Stop	GIS with stop locations.	100%

Note: Items in *italics* are optional.

(scheduled running time plus recovery time) should be based on an extreme value such as the 95th-percentile running time. However, without enough data to estimate the 95th-percentile running time, traditional practice sets it equal to a fixed percentage (e.g., 15% or 20%) of scheduled running time. Yet, some route-period combinations need more than this standard, and others less, because they do not have the same running time variability. AVL data allows an agency to actually measure 95th-percentile running times and use that to set recovery times. A study for Tri-Met found that basing recovery times on 95th-percentile running times would lead to substantial changes in cycle time that, compared to their current schedule, could reduce annual operating cost by an estimated $7 million (*28*).

- **Passenger waiting time** is an important measure of service quality. Studies show that customers are more affected by their 95th-percentile waiting time—for a daily traveler, roughly the largest amount they had to wait in the previous month—than their mean waiting time, because 95th-percentile waiting time is what passengers have to budget in their travel plans to be reasonably certain of arriving on time.
- **Passenger crowding** is also a measure in which extreme values are more important than mean values. Although traditional planning uses mean load at the peak point to set headways and monitor crowding, planners understand that what matters for both passengers and smooth operations is not mean load but how often buses are overcrowded. Therefore, design standards for average peak load are set a considerable margin below the overcrowding threshold. However, load variability is not the same on every route. With a large sample of load measurements, headways can be designed and passenger crowding measured based on 90th-percentile loads, or a similar extreme value, rather than mean loads.

4.1.2 Customer-Oriented Service Standards and Schedules

AVL-APC data allows customer-oriented service quality measures to replace (or supplement) operations-oriented service standards. For example, on high-frequency routes, a traditional operations-oriented standard of service quality is the coefficient of variation (*cv*) in headway. Although such a standard may mean something to service analysts, it means nothing to passengers, and it resists being given a value to passengers (e.g., how much does it benefit passengers if the headway *cv* falls from 0.35 to 0.25?). With a large sample size of headway data, one can instead measure the percentage of passengers waiting longer than *x* minutes, where *x* is a threshold of unacceptability. Similarly, in place of average load factor as a crowding standard, one could use a standard such as "no more than 5% of our customers should experience a bus whose load exceeds *x* passengers."

As these examples show, a shift toward customer-oriented measures goes hand-in-hand with the ability to measure extreme values.

4.1.3 Planning for Operational Control

One of the questions posed by the explosion of information technology is how best to use information in real time to control operations, for example, by taking actions such as holding a bus to protect a connection or having a bus turn back early or run express. As agencies experiment with, or use, such actions, they need off-line tools to study the impacts of these control actions in order to improve control practices. For example, AVL-APC data were used to determine the impacts of a Tri-Met experiment in which buses were short-turned to regularize headways during the afternoon peak in the downtown area (*29*).

4.1.4 Solutions to Roadway Congestion

Transit agencies are more actively seeking solutions to traffic congestion, such as signal priority and various traffic management schemes. They need tools to monitor whether countermeasures are effective. For example, a Portland State University study done for Tri-Met using archived AVL-APC data found that while signal priority reduced running time on some routes, it had no positive effect on others (*30*). In that particular study, only the overall effect on rather long segments was analyzed by comparing before and after running times, making the results hard to correlate with particular intersections. For better diagnosis and fine-tuning of countermeasures, agencies need tools to analyze delays on stop-to-stop, or shorter, segments.

4.1.5 Discovery of Hidden Trends

Behind a lot of the randomness in transit operations may be some systematic trends that can be discovered only with large data samples. For example, by comparing operators with others running the same routes in the same periods of the day, Tri-Met found that much of the observed variability in running time and schedule deviation is in fact systematic: some operators are slower and some faster. Exploratory analysis might also reveal relationships that can lead to better end-of-line identification, or to better understanding of terminal circulation needs.

It is the nature of exploratory analysis to not have a predefined format. To support exploratory analysis, therefore, AVL-APC databases need to be open to standard data analysis tools.

4.2 Key Dimensions of Data Needs

Along with listing uses of AVL-APC data, this chapter seeks to identify the particular data needs for each use, so that people involved in AVL-APC system design can better determine what features are needed to support various analyses. As indicated in Table 4, the needs of various analysis tools can be examined along three dimensions: basic record type, data detail, and sampling rate.

4.2.1 Basic Record Type

Four basic record types, giving vehicle location at regular intervals, are considered. (For more detail on record types, see Sections 2.3.1 and 2.3.2.) Polling records are not suited to stop or timepoint matching and therefore are only suitable for manual investigations involving playback. Timepoint records and stop records differ chiefly in level of geographic detail. In addition, stop records, which may or may not include passenger counts, are assumed to include the time at which doors opened and closed, or time of passing a stop if a bus does not stop. Timepoint records are assumed only to have either arrival or departure times. Interstop records are records of speed between stops; the term can also refer to summary data about what occurred between stops that may be part of a stop record.

4.2.2 Data Detail

This second dimension indicates what additional data are needed, either as additional items in the basic record type or captured in infrequent event records. Data detail is mainly affected by what devices are integrated into the AVL system.

4.2.3 Sampling Rate

As mentioned earlier, analyses involving estimation of mean values require a relatively small sample, while estimating extreme values or proportions requires large sample sizes. Therefore, it makes sense, and is consistent with practice, to treat this dimension as binary: either 100% (all vehicles equipped) or 10% (a sample of the fleet equipped), noting however that some uses that demand large sample sizes on a subset of routes can be accommodated without instrumenting the entire fleet, if the instrumented percentage is managed carefully.

4.3 Targeted Investigations

Analysis tools may be used in targeted investigations, which may be conducted to support legal, payroll, operations, maintenance, and other functions. For this category, transit agencies use archived AVL data to investigate suspected operator misbehavior and customer complaints, incidents, or accidents. These cases can be investigated using playback (a capability of many AVL systems oriented to real-time applications) in which data from the subject trip is viewed as if it were happening in real time. If stop or timepoint records are stored in a database, cases can be investigated more efficiently using general query capability.

For investigating incidents, it is often helpful simply to determine whether the bus in question was really there. Several transit agencies report having refuted accident claims by referring to archived AVL data. GPS-based location systems can indicate whether a bus was off route. Being able to relate the AVL data to operator-initiated event records and to control messages sent by dispatchers can be helpful for some types of investigations.

With interstop records, speed and possibly acceleration information can be extracted from an AVL database; however, to date, no agencies are known to sample so frequently as to give their AVL system the "black box" function of trip recorders that have become common for accident investigations in aviation and trucking.

For investigating customer complaints, simple playback (or better yet, a database query) can identify whether a bus was very early or late, or (with GPS data) off route. However, because being early is often a matter of only 1 or 2 min, timepoint records are better than polling records for verifying complaints about early buses. Also, because an early arrival does not necessarily mean an early departure, the location data should indicate departure time.

Some agencies use their AVL data to investigate operator overtime claims. AVL data has shown, for example, that an operator who reported late to the garage actually finished his last trip on time, suggesting that the late pull-in was intentional. Speed data from interstop records can be used to monitor speeding; however, experience has shown that speed data can be quite unreliable.

4.4 Running Time

Analyzing and scheduling running time is one of the richest application areas for archived AVL-APC data. Without AVL data, agencies must set running times based on small manual samples, which simply cannot account for the running time variability that comes with traffic congestion.

Buses are scheduled at the timepoint level; therefore, scheduling demands timepoint data. Because schedules sometimes refer to arrivals as well as departures, it is helpful if timepoint records include both arrival and departure times.

Running time analyses that require only estimation of mean values, or that involve only occasional studies (e.g., delay and dwell time analysis), can be conducted with only a sample of the fleet equipped with AVL. However, routine scheduling applications based on extreme values need the entire fleet equipped.

As part of this project, a set of improved analysis tools was developed for analyzing running time. They are mentioned in this section, but illustrated and further described in Chapter 5.

4.4.1 Allowed Time, Half-Cycle Time, and Recovery Time

A common analysis examines the distribution of observed running time for scheduled trips across the day compared with scheduled running time, also called "allowed time." An example is given in Figure 2, where the vertical bars show mean observed running time, the short lines show 85th-percentile values, and the arrows indicate maximum observed running times. Heavy horizontal lines show scheduled running time. This figure also distinguishes net and gross trip time, as explained in Section 4.4.4. When the number of observations is not too large, a scatterplot showing every observation can be useful.

Based on the observed distribution of running time for either a single scheduled trip or a set of contiguous trips in a period that will be scheduled as a group, schedule makers can choose a value for allowed time according to their preferred scheduling philosophy. Some schedule makers prefer to base schedules on mean running time. An alternative approach, aimed at improving schedule adherence, is to intentionally put slack into the schedule; this approach has to be coupled with an operating practice of holding at timepoints. With such a schedule, a high percentage of trips depart almost exactly on schedule, and the low percentage of trips that run late are not far behind schedule.

The amount of slack put into a schedule is often a simple fraction of mean running time, with ad hoc adjustments based on experience. A more scientific, data-driven approach is to use a percentile value, or "feasibility criterion." To illustrate, a feasibility criterion of 85% means setting allowed time equal to 85th-percentile observed running time; such a schedule can be completed on time 85% of the time. This approach is used in Chapter 5, with both the viewpoint of design ("tell me the 85th-percentile running time") and analysis ("tell me what feasibility I'd get if I added 1 minute to allowed time").

Analysis of running time is also pertinent for determining how much recovery time to schedule at the end of the line. The time from a bus's departure at one terminal to its next departure in the reverse direction has been called the "half-cycle time"; it is the sum of running time and recovery time. Because the purpose of recovery time is to limit the likelihood that delays encountered in one trip will propagate to the next, half-cycle time is based logically on a high-percentile value of running time. Tri-Met has begun to systematically revise its half-cycle times, basing them on a 95% feasibility criterion so that there will be only a 5% chance that a bus will arrive so late that it starts the next trip late. For this applica-

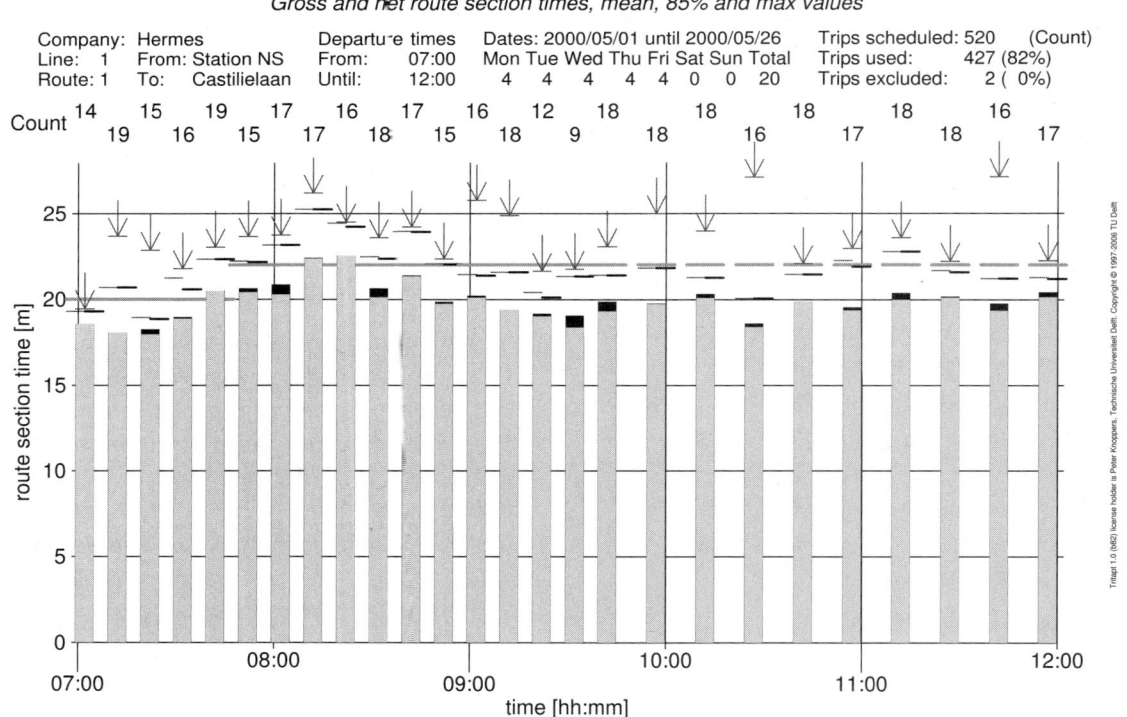

Figure 2. Observed running time by scheduled trip.

tion, scheduled recovery time is set to be the difference between 95th-percentile running time and allowed time.

The chief engineer for operations analysis at Brussels' transit agency has recently been working with a scheduling software vendor to develop reports to support statistically based schedules. Their approach uses three parameters, as explained in this example: If the three parameters are 95%, 80%, and 5 min, then half-cycle time is set equal to 95th-percentile running time, and allowed time is set equal to 5 min less than 80th-percentile running time. That way, there will be an 80% chance that a trip finishes no more than 5 min late, and a 95% chance that the next trip can start on time.

4.4.2 Segment Running Time

Scheduling running time on segments, or equivalently setting departure (or arrival) times at timepoints relative to the trip start time, can either precede or follow scheduling route times. One approach, common in U.S. practice, is to first determine segment running time based on mean observed running time by segment, perhaps adding a certain percentage for slack, and then constructing route time as simply the aggregation over the segments.

Simply aggregating over segments will not work with a feasibility-based approach, because the sum of the parts will not yield a valid measure for the whole. For example, the sum of 85th-percentile segment running times does not equal a route's 85th-percentile running time; the sum will be far greater, in fact.

The Delft University of Technology has developed the Passing Moments method of extending statistically based scheduling to the segment level (*31*). Applied at several Dutch transit agencies, the Passing Moments method bases timepoint schedules on f-percentile completion time from each timepoint to the end of the line, where completion time is running time from a point to the end of the line, and f is the feasibility criterion (e.g., 85%). Segment running times are determined by working backwards from the end of the line, without ever explicitly analyzing observed running time on timepoint-to-timepoint segments. This approach was designed to overcome operators' resistance to holding, which is the key to good schedule adherence. If a schedule is written based on mean running time, operators know that, if they hold at a timepoint, they will have a 50% chance of finishing the trip late and thereby getting a shortened break; therefore, they are reluctant to hold. With 85th-percentile allowed times between each timepoint and the end of the line, operators know that even if they hold, they have a high chance of finishing on time.

The Passing Moments method is one of the scheduling tools described in Chapter 5. The Passing Moments method has two advantages compared to setting slack time simply proportional to mean running time. First, it is sensitive to where on the route delays and running time variability occur. Second, it tends to put less slack in the early part of the route and more in the later part of the route, which is a better way to distribute slack than simply applying it proportionately throughout. It results in holding buses less in the early part of the route because they may need that time later in the route.

Steve Callas, the manager of Tri-Met's AVL data analysis program, has suggested that if operating practice is such that buses do not hold at timepoints, it may be better to use a low feasibility criterion because running early is more harmful to passengers than running late. For example, a statistically based approach for segment-level scheduling might be to base departure time at a timepoint on 40th-percentile observed cumulative running time (i.e., from the start of the line to a point). That way, there will be only a 40% chance of a bus departing early, and those that depart early should not be very far ahead of schedule.

4.4.3 Choosing Homogeneous Running Time Periods

Another problem in running time analysis is choosing the boundaries of running time periods within which allowed time is constant. Establishing periods of homogeneous running time involves a trade-off between short periods within which scheduled running times match the data well versus longer periods of constant allowed time but greater variability. A common logic for resolving this trade-off is first to determine, for each scheduled trip, an ideal allowed time (e.g., mean running time, or 85th-percentile running time, depending on the desired feasibility criterion) and then to make running time periods as long as possible subject to the restriction that no more than a certain percentage of the scheduled trips in that period have an ideal running time that deviates by more than given tolerance from the suggested running time for that period.

4.4.4 Excluding Holding (Control) Time

Ideally, scheduling tools should use net running time, which excludes holding time, also called "control time." Identifying what part of observed running time is holding time can be tricky, requiring greater data detail, and is done by only a few agencies with AVL-APC data. Ideally, stop records should indicate both when doors open and close, and when the wheels start to roll, something provided by at least some APC vendors. If the bus is ahead of schedule, any unusual gap between door close time and departure time can be interpreted as holding.

The running time analysis shown earlier in Figure 2 distinguishes gross from net running time. The gray bars show net running time; their black tops are control time, making the combined height equal to gross running time. Of the short horizontal lines representing 85th-percentile values, those extending to the left of a scheduled start time are gross running times, and those extending to the right are net running times.

Because in good weather holding may occur with the doors open, NJ Transit's system is being upgraded to provide a record of how many passengers boarded and alighted every few seconds (the main on-board computer frequently queries the on-board APC analyzer while doors are open and writes records), allowing NJ Transit to recognize periods of inactivity even while the doors are open. Eindhoven's algorithm looks for unusually long dwell times when the bus is ahead of schedule.

Event codes for lift use or a fare dispute can also help identify holding time by explaining the cause for long dwell times.

Another version of holding is still harder to detect—"killing time" en route to avoid being early. Data and algorithms that would help detect "killing time" would be valuable and are being developed at NJ Transit.

4.4.5 Stop-Level Scheduling

In many European countries including the Netherlands, schedules are written at the stop level. In fact, on many routes, every stop is a timepoint. Making (almost) every stop a timepoint has the advantage of replacing occasional large holding actions with frequent small ones, which are less obvious and irritating to passengers.

Stop-level schedules fit well with the trend of giving customers better information. Stops are where customers meet the system, and where they need to know scheduled departure times. Internet-based trip planners need stop-level schedules, as do real-time next-arrival systems (to know stop-to-stop expected running time). In current practice, agencies with customer information systems like trip planners and real-time information systems estimate stop-level departure and running time by interpolation between timepoints; using stop-level AVL data to develop stop-level schedules offers an obvious improvement. Not providing the public with stop-level schedules is a good example of practice being driven by the historic lack of data—practice that should change as advanced technology is deployed in transit.

Stop-level schedules are valuable for control even when every stop is not a timepoint. Some AVL systems display schedule deviation to the operator, who can use this information all along the route to try to adjust bus speed. For example, in Eindhoven a small display for operators shows schedule deviation in units of 10 s. Such a system can be effective, however, only if it is based on a realistic, finely tuned schedule. Simply interpolating between timepoints is too approximate if speed between timepoints is not uniform, such as when the route passes major intersections. For operators to have confidence in, and therefore use and benefit from, a schedule deviation display, they need data-based stop-level schedules.

Signal priority is another application that needs stop-level schedules, if priority is conditional on buses being late. Conditional priority is a form of operational control, pushing late buses ahead while holding early buses back, and to be effective, it needs a finely tuned schedule at every signalized intersection. This system has been used very effectively in Eindhoven (*32*). Conditional priority without fine-tuned schedules can easily devolve into either unconditional priority (because buses always arrive late) or no priority (because buses arrive early). To have a good control margin, schedules have to be written so that the probability of arriving late is far from the extremes of 0% and 100%.

4.4.6 Speed and Delay Analysis

Speed, delay, and dwell time studies are analyses that help support a transit agency's efforts to improve commercial speed, something that benefits both operations and passengers. "Speed" in this context is average speed over a segment, not instantaneous or peak speed. A display such as given in Figure 3 showing delay by segment (or, alternatively, average speed by segment) helps a transit agency to identify problem locations, to monitor the impacts of actions that affect speed, and to monitor and document historic trends in operating speed. In that figure, the thin horizontal lines are individual observations of delay by segment; the box height is the 85th-percentile delay, and the bar inside the box indicates mean delay. Analysts will be interested not only in average delay, but also in how variable it is, and in the likelihood of extreme values.

A report showing delays or speeds between stops offers a richer, more geographically detailed view than one using timepoint segments. Another reason to prefer stop records as the basis of delay analysis is that it allows dwell time and control time (which almost always occur at stops) to be removed, which puts a clearer focus on the effects of the roadway and traffic on bus speed and delay.

"Delay" can be defined in several ways. Two definitions of delay are (1) the travel time between stops minus the average travel time measured during non-congested periods such as early morning or late evening and (2) the amount of time spent at speeds below 5 km/h minus the time spent at stops. (Eindhoven uses this second definition.) To support this definition of delay, an AVL system needs records of when speed thresholds are crossed.

4.4.7 Dwell Time Analysis

Transit agencies also try to improve commercial speed by reducing dwell time, using such measures as low-floor buses or changes to fare collection equipment and practices. Stop records with door open and close times allow agencies to analyze dwell time to determine impacts and trends.

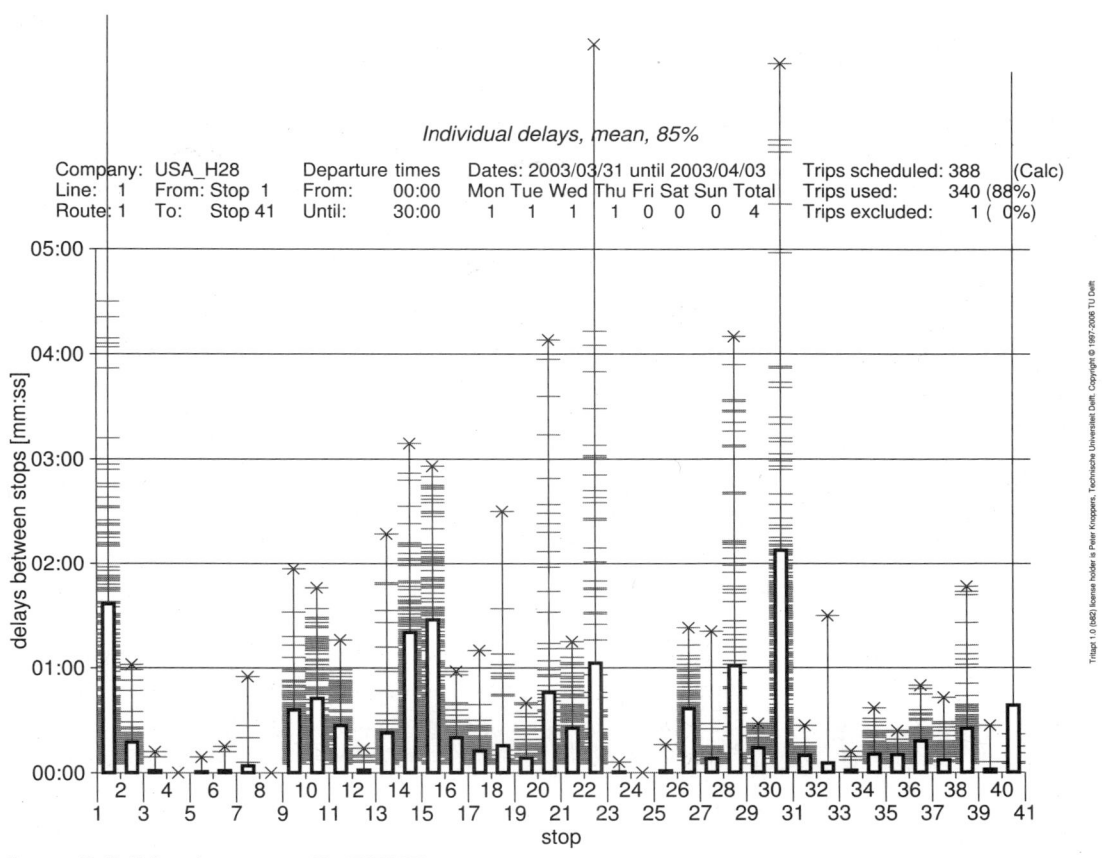

Figure 3. Delays by segment.

Such an analysis should preferably be aided by passenger counts, in order to separate out the impact of the number of boardings and alightings and to identify whether any on-vehicle congestion impact arises when vehicles are crowded. On-off counts, farebox transactions, and incident codes that reveal wheelchair and bicycle use are all useful for giving analysts an understanding of dwell time.

4.5 Schedule Adherence, Long-Headway Waiting, and Connection Protection

Monitoring schedule adherence is a valuable management tool, because good schedule adherence demands both realistic schedules and good operational control. It is probably the most common analysis performed with AVL-APC data.

Schedule adherence can be measured in a summary fashion as simply the percentage of departures that were in a defined on-time window, or perhaps as the percentage that were early, on time, and late. Standard deviation of schedule deviation is an indicator of how unpredictable and out of control an operation is; along with schedule adherence, it is part of a daily service quality report in Eindhoven.

A distribution of schedule deviations provides full detail. Such a distribution allows analysts to vary the "early" and "late" threshold depending on the application, or to determine the percentage of trips with different degrees of lateness.

A profile of schedule deviations along the line is a valuable tool, showing how both the mean and spread of schedule deviation changes from stop to stop. Figure 4 shows two examples taken from Eindhoven. The heavy, black line indicates mean schedule deviation; the heavy, gray lines indicate 15th- and 85th-percentile deviation. Thin lines represent individual observed trips. How close the mean deviation is to zero indicates whether the scheduled running time is realistic. If the mean deviation suddenly jumps, it means the allowed segment time is unrealistic. Deviations at the start of the line are particularly informative: if most trips are starting late, it might indicate that the route's allowed time is too long, and that operators are starting late to avoid running early. The spread in deviations, and how much it increases along the line, is a good indicator of operational control. The display in Figure 4(a) shows a poorly scheduled and poorly controlled route; Figure 4(b), in contrast, shows a route for which most schedule deviations remain in the 0- to 2-min band all along the line.

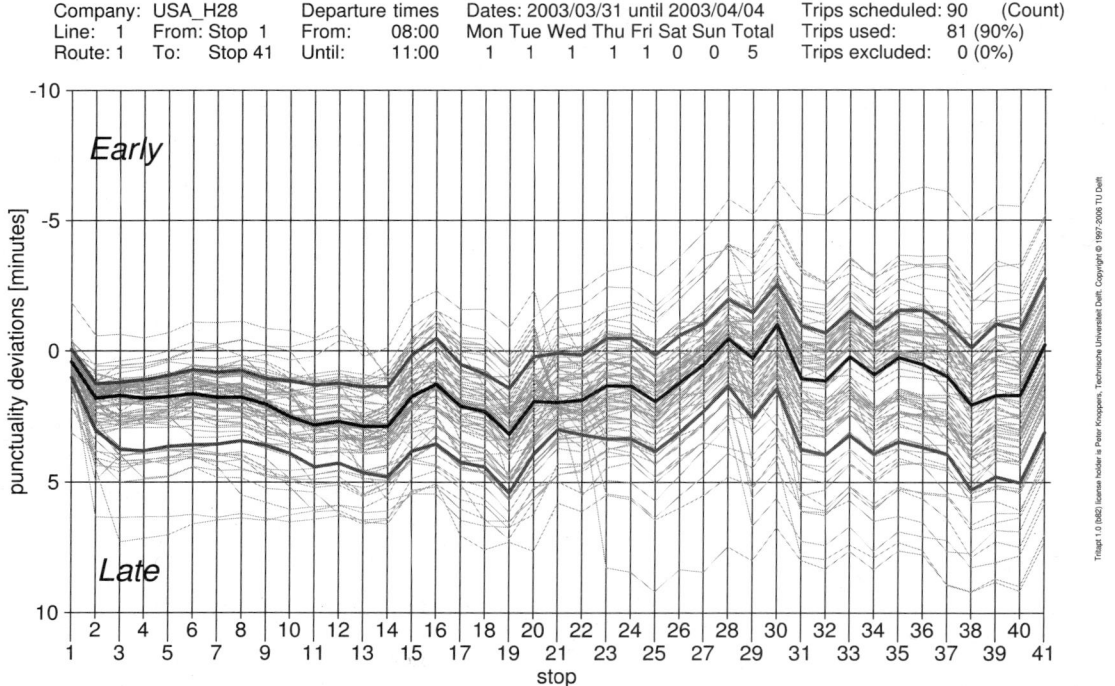

(a) Showing Both Systematic and Strong Random Deviation

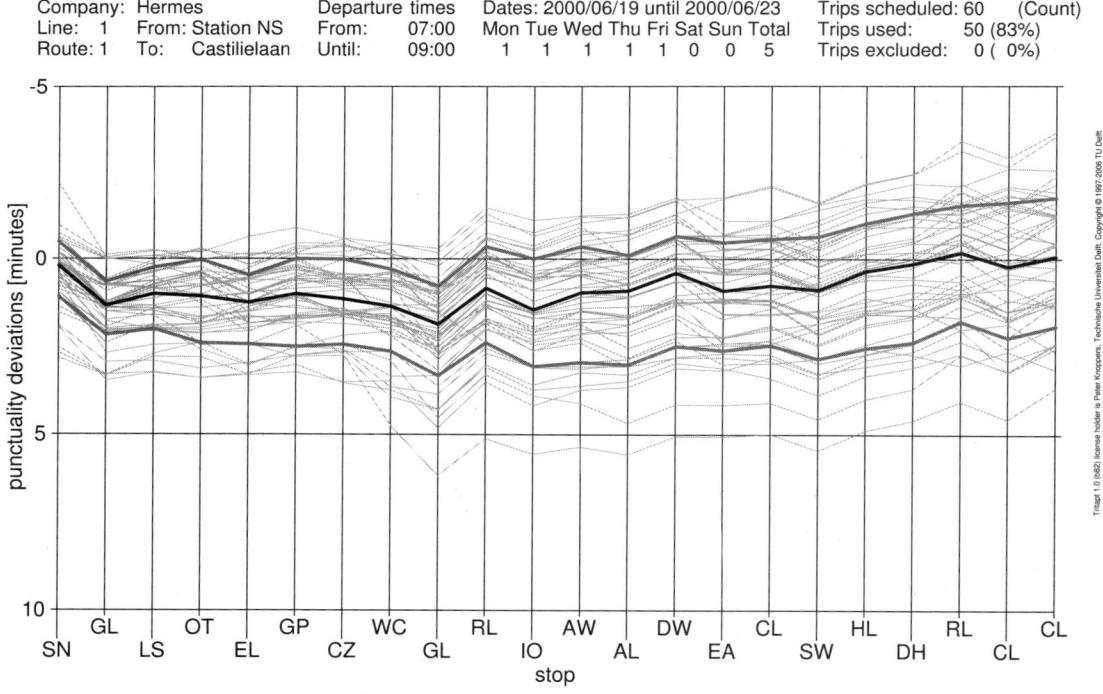

(b) Showing Little Systematic or Random Deviation

Source: Hermes (Eindhoven), generated by TriTAPT

Figure 4. Schedule deviation along a route.

Because schedules are written at the timepoint level, timepoint data will support schedule adherence analysis. And because schedule adherence involves estimating proportions and extremes (detecting the percentage of early and late trips), the full fleet should be equipped. Finally, because schedules sometimes refer to arrival time as well as departure time, a data collection system that captures both is preferred.

Passenger waiting time on routes with long headways is closely related to schedule adherence. Chapter 6 shows how it is possible to determine excess waiting time from the spread between the 2nd-percentile and 95th-percentile schedule deviation.

Passengers are particularly interested in whether they can make their connections. Arriving 4 min late is not a problem if the time allowed for the transfer is 5 min, but it could be a big problem if the allowed time is only 3 min. However, if the departing trip is held—again, the convergence of schedule planning and operations control—other issues arise. AVL data is ideal for determining whether specific connections were met. To analyze connection protection an agency must define the particular connections it wishes to protect or at least analyze. The researchers found one transit agency using its archived AVL data for this purpose. Integrating control message data, which might include requests for holding to help a passenger make a connection, would permit a deeper analysis of operational control. Incorporating demand data, ideally transfer volumes, would make the analysis richer still.

Connection protection analysis requires data structures and software that create the capacity to perform analyses across routes.

4.6 Headway Regularity and Short-Headway Waiting

On routes with short headways, headway regularity is important to passengers because of its impact on waiting time and crowding. It is also important to the service provider because crowding tends to slow operations and because much of operations control is focused on keeping headways regular.

To measure headways, data has to be captured on successive trips, making headway analysis particularly sensitive to the rate of data recovery, as one lost trip means two lost headways. Analyzing headway when only part of the bus fleet is instrumented poses the logistical challenge of getting all the buses operating on a route to be instrumented; because of this challenge, Table 4 indicates headway analysis needs 100% of the fleet to be instrumented with AVL.

Headways matter all along the route, not only at timepoints; therefore, stop records are best suited to headway analysis. (In fact, headways matter most at stops with high boarding rates.) However, because headways at neighboring stops are strongly correlated, timepoints can be thought of as a representative sample of stops, making it possible, although not ideal, to estimate headway-related measures of operational quality from timepoint data. To the degree that operators hold at timepoints, however, using them as representative stops becomes deceptive.

4.6.1 Plotting Trajectories

One much-appreciated AVL data analysis tool is a plot of successive trajectories on a route, as illustrated in Figure 5. Its format shows observed versus scheduled trajectories for a line-direction. In the color version of this graph, each bus appears in a different color so that bus-specific trends (e.g., a slow driver) can be spotted. This kind of analysis is helpful for illustrating the dynamics of bunching and overtaking and for showing where delays begin and how they propagate. However, while this tool is helpful for giving the sense of how a route operates, it does not yield any numerical results and is suitable only for analyzing a single day's data at a time.

4.6.2 Headway Analysis

A numerical analysis of headway data applies over multiple days, for a route-direction and a period of the day with relatively uniform headways. Typical summary results are mean and coefficient of variation (*cv*, which is standard deviation divided by mean) of headway. On short-headway routes, the *Transit Capacity and Quality of Service Manual* (*27*) assigns levels of service for service reliability based on values of headway *cv*. Mean headway can be compared with mean scheduled headway to see whether more or less service than scheduled was operated. In place of headway *cv*, Eindhoven uses a regularity index, which is the mean value of the absolute headway deviation divided by mean headway.

A distribution of headways is an even richer result than mean and *cv* of headway, allowing analysts to see how often headways were very short or very long, using any threshold they desire. For its rapid transit routes, the MBTA uses the percentage of headways greater than 1.5 scheduled headways as an indicator of service quality.

Analysis procedures have to be careful in dealing with period boundaries. To illustrate, if the morning peak period ends at 9:00, a trip scheduled to pass a timepoint at 8:58 may pass on some days before 9:00 and on other days after 9:00. In an analysis of headways in a period ending at 9:00, that trip will sometimes be counted and sometimes not, introducing variability into the analysis as an artifact.

4.6.3 Passenger Waiting Time

On short-headway routes, passengers can be assumed to arrive at random; therefore, passenger waiting time can be

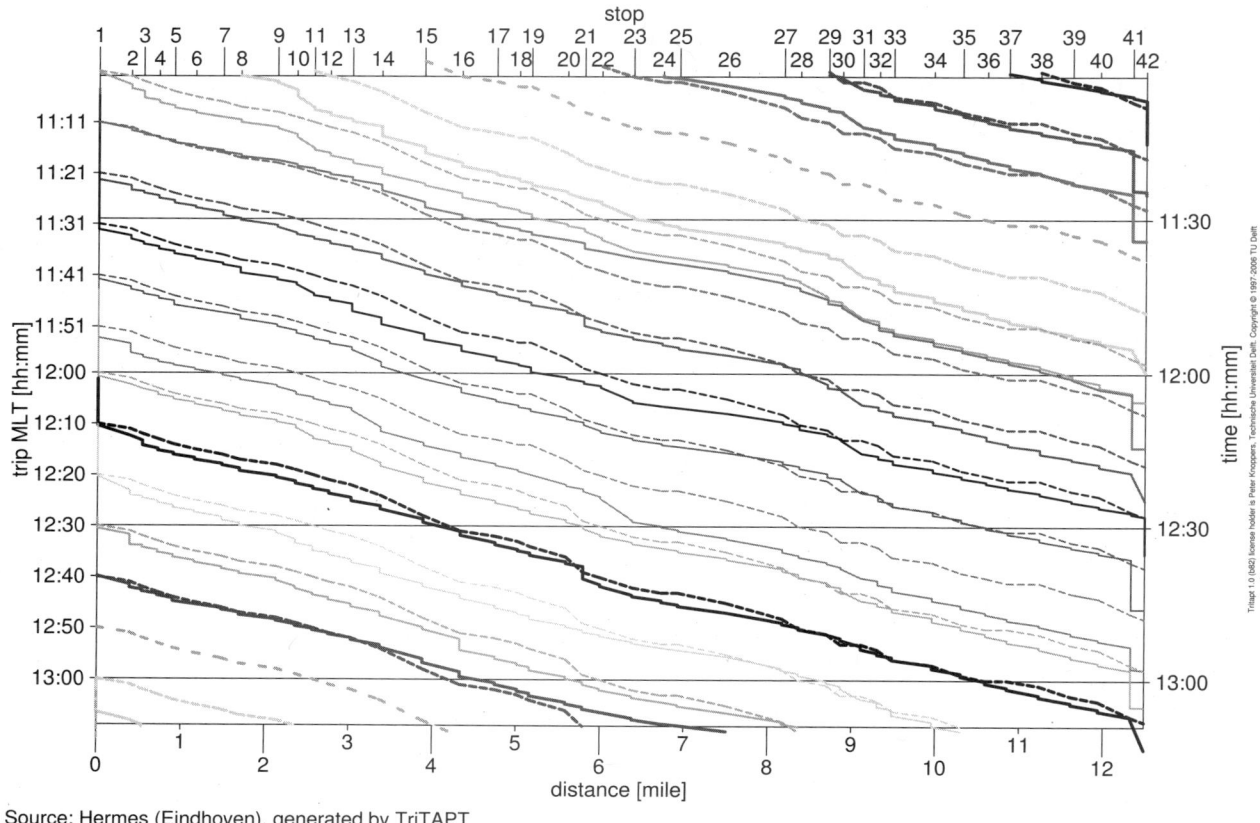

Figure 5. *Observed (solid) versus scheduled (dotted) trajectories.*

determined and analyzed directly from the headway distribution. As an example, transit agencies in both Brussels and Paris calculate, from headway data, the percentage of passengers waiting longer than the scheduled headway plus 2 min. As part of this project, analyses of passenger waiting time based on headway data were developed (see Chapter 6).

4.6.4 Headway-Load Analysis

Headway and load have a simultaneous effect on each other—longer headways lead to larger loads, and larger loads lead to longer headways. Analyzing headway and load together can lead to interesting insights.

By analyzing headway and load together, Tri-Met created a method for determining to what extent an overload is caused by headway variation, as opposed to demand variability (28). The idea behind this approach is that overloads caused by headway variation should be "cured" by better headway control, while overloads that cannot be simply explained by headway variation may require a change in scheduled departure times or headways. Tri-Met estimates, for a given route-direction-period, the slope of the headway-load relationship using a least-squared fit. Then, using that slope, Tri-Met normalizes the loads of individual trips to what the loads would have been if the headway had been as scheduled.

4.7 Demand Analysis

The sources of passenger use data are APCs and fare collection systems. In this report, the only fare records considered are location-stamped transactions, because analysis of farebox data at the route level or higher is routine.

4.7.1 Demand Along a Route

Passenger demand on a route, for a given direction and period of the day, has three dimensions: geographic (i.e., along the route); between scheduled trips (i.e., how is the 7:15 trip different from the 7:30 trip); and between days. Most analysis views aggregate over two dimensions and analyze the remaining one; it is also possible to aggregate over only one dimension, showing the other two in the analysis, as the examples will show.

The geographic dimension of demand is shown in a volume profile or load profile, depending on whether results are expressed in passengers per hour (volume) or passengers per trip (load); another view shows ons and offs by stop. One graphical format, developed by Delft University researchers and illustrated as the lower step line in Figure 6, shows not only mean segment loads, but also mean offs, ons, and through load at each stop in a single profile. The upper, gray step function indicates 85th-percentile segment loads, thus adding the dimension of day-to-day variation. This report has already pointed out the importance of extreme values of load for both passenger service quality monitoring and scheduling. Also shown in Figure 6 for each stop, as well as for the route as a whole, are box and whiskers plots of offs (just to the left of each stop) and ons. The box extends from the 15th percentile to the 85th percentile; a bar indicates the mean value; and the X above the box indicates the maximum observed value.

Analysis of demand along a route is necessary for understanding where along the route high loads occur. It supports decisions about stop relocation and installing stop amenities, and routing and scheduling actions that affect some parts of a route differently from others, such as short turning, zonal service, and limited stop service (*33*).

4.7.2 Demand Across the Day and Scheduling Headway and Departure Time

By abstracting the geographical dimension using trip summary measures such as total boardings, maximum load, and passenger-miles, one can focus on the other two dimensions of demand, variation within a day and between days. Figure 7 shows how demand varies between scheduled trips, with scheduled trips on the horizontal axis and one measure of demand, in this case mean boardings, on the vertical. In other versions of this graph (not shown), day-to-day variation is presented by showing a scatterplot (horizontal whiskers) or selected percentile values, which allows one to see extreme values of load that are important to both scheduling and operational control. Using established thresholds, trips can be categorized and counted by degree of crowding.

In Figure 7, four of the scheduled trips in the period analyzed had no valid APC data. They are represented with a large X and a more darkly colored bar whose height is set equal to the average of the nearest trip before and after it with valid counts. The issue of imputing values to missing data is discussed in Chapter 11.

Passenger-miles is another summary measure over a route, being the product of the segment load multiplied by the

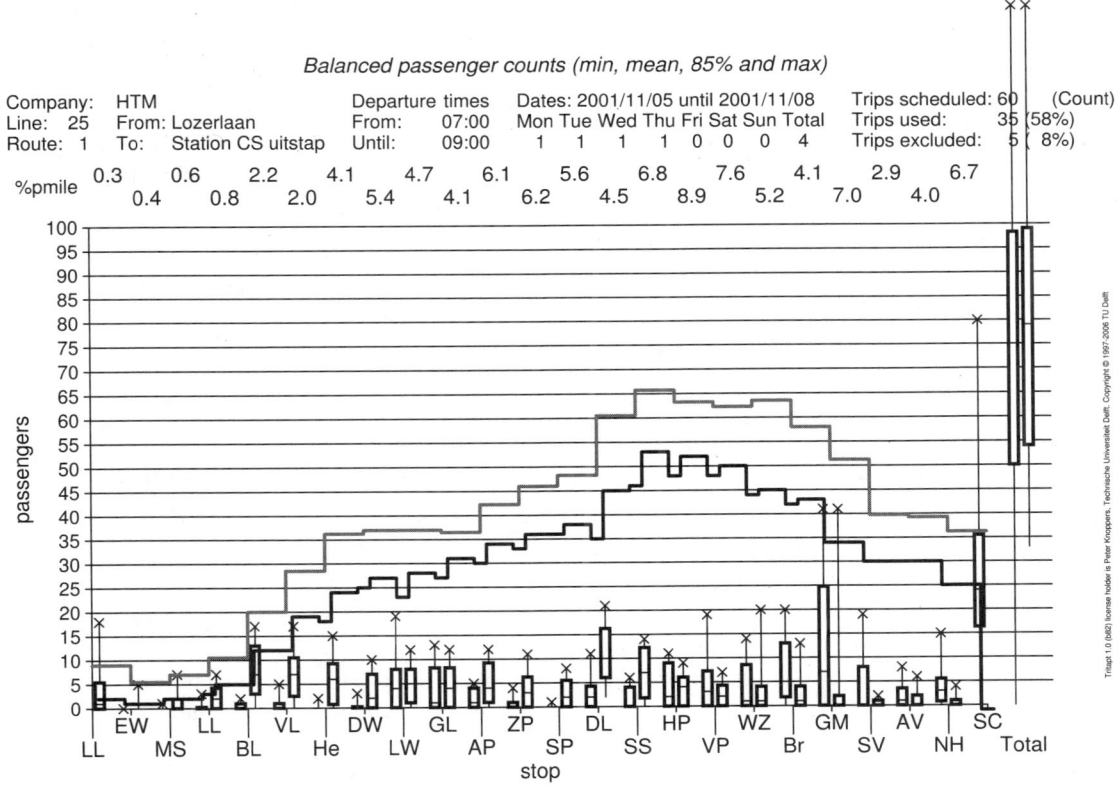

Figure 6. Load and on/off profile.

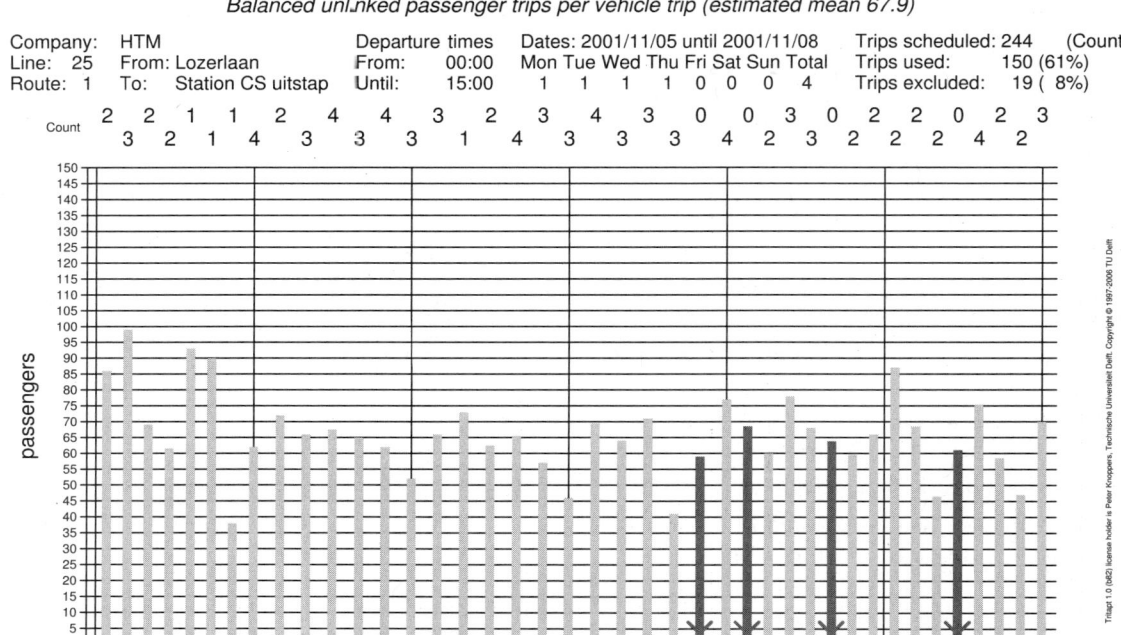

Figure 7. Boardings by trip across the day.

segment length. When divided by overall route length, this total indicates the average vehicle occupancy along the route. Special considerations relative to measuring passenger-miles are covered in Chapters 8 and 9.

An analysis of demand variation across the day supports scheduling, which, in part, sets headways and departure times so as to achieve target loads. There remains the opportunity to develop design tools for scheduling that take advantage of large APC sample sizes to estimate a demand profile across the day. Using passenger counts combined with measured headways, and averaging over many days, one should be able to derive the passenger arrival rate as a function of time. Combining these arrival rates in small (e.g., 1-min) time slices using a reference frame that moves at the speed of a bus allows one to predict the peak load on a trip based on its departure time and the departure time of its leader.

With a minute-by-minute load profile across the day, one valuable tool would be able to find periods of homogeneous demand within which a constant headway can be used, analogous to scheduling tools that seek periods of homogeneous running times. Another valuable tool would not assume constant headways at all, but would select departure times that balance loads between trips, accounting for how demand rates vary across the day, as suggested by Ceder (34). Tools of this sort are currently under development for the transit agency of the Hague.

In the future, there may be scheduling tools that account for within-day and between-day variation in demand, as well as within-day and between-day variation in running time, in order to design route schedules that respond to how both demand and running times vary across the day, using statistical methods to limit the probability of overcrowding and insufficient recovery time.

4.7.3 Passenger Crowding

There is a strong relationship between vehicle crowding and passengers' experience of crowding, but the perspectives are different. For example, if half the trips are empty and half are overcrowded, then only 50% of the trips are overcrowded, yet 100% of the passengers experience an overcrowded trip. Measures of crowding from the passenger perspective are discussed in Chapter 7.

4.7.4 Pass-Ups and Special Uses

Operator-initiated incident codes used to register such events as pass-ups, wheelchair customers, and bicycle customers can be used to analyze special demands and events along a route or across the day. Being able to locate them along a route might be useful for load analysis, running time analysis, and facility planning.

4.8 Mapping

Equipped vehicles can serve as GPS probes whose archived AVL data is used to improve a transit agency's base map. For example, if buses stop often at a location not indicated on the base map as a stop, then a stop may be missing on the base map (perhaps because it has been informally added by operators); this data can be used to help locate both permanent and temporary stops.

A more explicit use of buses as GPS probes is to intentionally use them to map a bus's path through a new shopping center or subdivision. For this application, the on-board computer has to be set to make frequent interstop records. An Israeli APC supplier includes a learning mode that allows an on-board surveyor, seated beside the operator and holding a laptop computer, to create geocoded records with codes and comments at points of interest (e.g., where a bus makes a turn) to help map the bus's path.

4.9 Miscellaneous Operations Analyses

The availability of archived AVL-APC data creates opportunities for analysis of many other aspects of operations, of which five are listed in Table 4 and discussed in this section. Other analysis opportunities will undoubtedly be discovered, highlighting the need for AVL-APC databases to support exploratory and new analyses.

4.9.1 Acceleration and Ride Smoothness

One aspect of service quality that might be measured with an advanced AVL system is the smoothness of the ride. Passengers value a smooth ride, without jerky accelerations or decelerations, while avoiding unsafe speeds. At present, transit agencies in Paris and Brussels use externally contracted surveyors called "mystery shoppers" to rate quality of service in several categories, including ride smoothness; their ratings are, of course, subjective. Very frequent records of speed would permit an objective measurement of linear speed, acceleration, and deceleration; swerving and bouncing also could be measured if accelerometers in three directions were integrated into the system.

4.9.2 Mechanical Demand

AVL data may permit analysts to estimate mechanical demands on buses in order to relate them to vehicle performance and maintenance. For example, combining measurements of vehicle acceleration and passenger load with GIS information on roadway grade allows estimation of the tractive and braking forces required, which then could be analyzed to find relationships to fuel consumption, brake wear, or engine maintenance needs. Another suggested measure of mechanical demand that could be determined from interstop AVL records is the number of acceleration/deceleration cycles.

4.9.3 Terminal Movements

Interstop GPS records might be used to analyze vehicle movements at terminals, which may be of interest at busy terminals with capacity, safety, or efficiency issues. A better understanding of terminal movements can also lead to better determination of arrival and departure times, which are critical for schedule analysis.

4.9.4 Control Messages

While operator-initiated messages (e.g., indicating pass-ups or bicycle use) are customarily coded in a manner that permits numerical analysis, control messages sent by radio to bus operators are not customarily so coded. To the extent they could be coded for common commands such as hold for the schedule or hold for a connection, they would allow one to analyze where and when those control messages are used, account for their impact on running time, and analyze their effectiveness.

4.9.5 Operator Performance

Finally, published (28) and unpublished studies by Tri-Met using AVL-APC data indicate that much of the variance in running time and schedule adherence can be explained by operator behavior. An analysis of performance by operator could be a valuable tool for training operators and for experimenting with different methods of supervision and control. To account for the bus bunching phenomenon, an operator's performance on short-headway routes should account for the position of its leader. Performance elements can include schedule deviation (especially at dispatch), running time, layover time, headway maintenance and bunching, and more.

Correlations between data items may reveal interesting operating patterns. Do operators that are beginning to run early intentionally slow down, and do operators that are getting behind speed up? Do operators drive differently when they have a heavy load or after they depart the terminal late? Being able to identify individual operators may reveal operator-specific patterns or relationships between running time and operator experience (both overall and on the specific route). Uncovering operating patterns like this can be useful for planning both schedules and methods of supervision and training.

Operator performance must be analyzed with careful respect for operator acceptance and safety. If used for discipline, data for such analyses may be in danger of sabotage. Agencies may not want to use the AVL data directly to discipline operators, but it can certainly be used to help dispatchers and supervisors better target their efforts at conventional discipline. For example, some agencies report that if data indicates a recurring problem with a particular trip starting late, a supervisor might be requested to observe.

More seriously, safety can be compromised if operators are punished for getting behind schedule. Such concerns do not necessarily mean that operators should not be given feedback on their performance. Experience with data collected automatically in Rotterdam's tram operation shows that operators may enjoy getting a written record of their performance—for the first time operators had written evidence to show their family what a good job they were doing in staying on schedule.

4.10 Higher Level Analyses

This section discusses analyses of AVL-APC data that cover extended periods of time or multiple routes.

4.10.1 Comparisons and Aggregations

By comparing results of analyses done over selected dates, AVL-APC data can be used in before–after studies or to analyze operations during special events or weather conditions.

Trends analysis can be seen as simply an extension of the before–after study, but it suggests a need for storing higher level summaries in a separate database. An example is a monthly systemwide report on schedule adherence. A transit agency might specify measures that it wants to follow over time, calculate those measures periodically (e.g., every month) from the detailed AVL-APC data archive, and save those period summaries in a smaller, higher level database where they can be used for trends analysis.

Many analyses that involve aggregation or comparison over routes can benefit from AVL-APC data. One example is a periodic route performance comparison, which may include items such as on-time performance or total boardings along with data from other sources such as scheduled vehicle-hours or farebox revenue. Another example is making annual systemwide passenger-miles estimates for reporting to the NTD, which can be made by aggregating mean passenger-miles per trip over all the trips in the schedule (see further discussion in Chapter 9).

These applications suggest having an automated process of periodically calculating and storing summary measures in higher level tables.

4.10.2 Transfers and Linked Trips

While APCs provide the data needed to analyze demand on a route, they do not capture the information needed to identify linked trips or transfers. However, if fare media include unique IDs (as is the case with both magnetic cards and smart cards), stop- and time-stamped farebox transactions permit analysis of transfers and linked trips. Linked-trip analysis is especially important in Canada, where linked trips are the standard measure of transit use.

Despite fare systems not capturing alightings by ID code, there have been successful efforts in New York, Chicago, and Dublin to determine transfers and linked-trip origins and destinations by tracking where a fare ID is next used to enter the system (*18, 19,* and unpublished work). This area is promising for future research.

4.10.3 Headways and Other Measures on Shared Routes

Many transit networks have trunks served by multiple lines or multiple patterns (branches) of a line. Some measures of activity on a trunk are simple aggregations of stop-level measures; examples are schedule adherence and passenger load. For these purposes, all that is needed is an interface allowing one to select the appropriate set of stops and patterns. However, headways on a shared route can only be determined by going back to original stop or timepoint records, including data from all the trips serving the trunk, and linking them where their respective route joins and leaves the common trunk. This procedure demands a special data structure for a route trunk, something developed as part of this project (see Section 11.5).

4.10.4 Geographic Analyses

Transit agencies often want to do route-independent analyses based on geography, including both demand analysis (how many boardings occur in a certain area) and service quality analysis (what is the on-time performance in a certain area). Integrating AVL-APC data with GIS models requires data structures that link geographic locations to stops and route segments, and a process to extract and aggregate results for the selected stops and segments.

For demand modeling, methods are needed to convert on-off counts at stops into trip generation rates in small traffic analysis zones. However, this specialized procedure could be driven equally by manual or automatically collected data; the challenge for APC data analysis is to export demand rates by stop for only a selected period of the day.

CHAPTER 5
Tools for Scheduling Running Time

This chapter describes running time analysis and scheduling tools that were developed and/or improved as part of this project. They use statistical methods to create running time schedules, taking advantage of the large sample sizes afforded by AVL data, and are part of TriTAPT software developed by researchers at the Delft University of Technology. The primary tools described in this chapter are packaged as two integrated analyses: the first divides the day into running time periods and establishes route running times for each period, and the second allocates running time over a route's segments. If the captured data allows the identification of control (holding) time, these running time tools will be applied to the net running time, which excludes control time.

Both analyses use graphical reports or screens, behind which are exportable tables generated from AVL data. They apply to a single route-direction, using data from any number of days.

5.1 Running Time Periods and Scheduled Running Time

The first analysis, called "homogeneous periods," is a semi-automated, interactive tool for establishing running time periods (periods of constant scheduled running time) and scheduled running times. This tool allows the user to examine the feasibility of the current set of scheduled running times or a user-proposed set of running times, and it also suggests running times and periods automatically.

5.1.1 Feasibility of the Current Timetable

Figure 8 shows an analysis of the current running times (vertical axis) and running time periods across the day (horizontal axis). Features include

- A statistical summary of observed running time for each scheduled trip, showing mean (gray bar height), two percentile values (jagged lines, set for this figure at 50th and 80th percentile), and maximum observed running time (arrow);
- Current running time periods, bounded by heavy vertical lines, with a similar heavy horizontal line indicating current allowed time; and
- Suggested allowed times (thick, gray horizontal lines), about which more will be said later.

At the bottom of the rectangle for each running time period is a calculated value called feasibility; it represents the percentage of observed trips in the running time period whose running time was less than or equal to the current allowed time.

5.1.2 Suggesting New Running Times and Running Time Periods

In the graph shown in Figure 9, a set of allowed times and running time periods suggested automatically by the program are shown and analyzed. A feasibility value is shown for each suggested period. Current allowed times are also visible in the background as solid horizontal line segments.

The algorithm that suggests running time periods and allowed times seeks a compromise between trying to closely match the data and having periods as long as possible in order to make scheduling and control simpler. TriTAPT offers users two algorithms for selecting homogeneous periods:

- For one algorithm, users set two percentile limits, for example, 50% and 80% (the values used in this section's figures). The algorithm then seeks periods for which a whole-minute running time can be suggested that lies between the 50th-percentile and 80th-percentile observed running time for (almost) every trip in that period.
- For the second algorithm, users specify a single feasibility value and a tolerance–for example, 85% and 2 min. Then, the algorithm seeks periods for which a running time can be

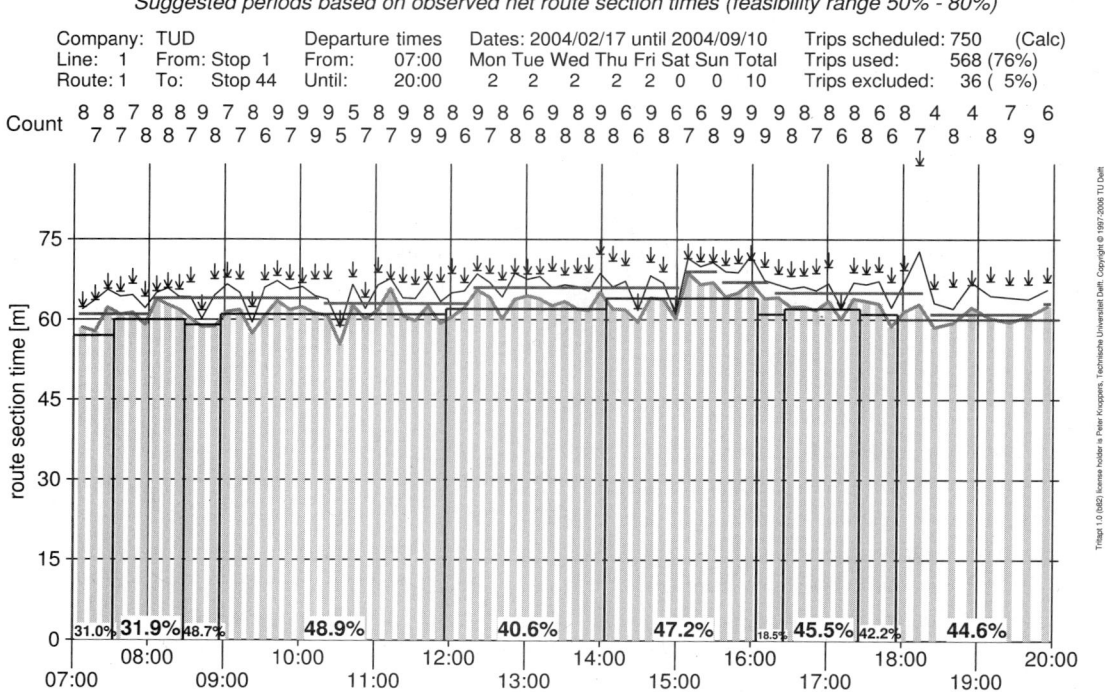

Figure 8. Analysis of current running times.

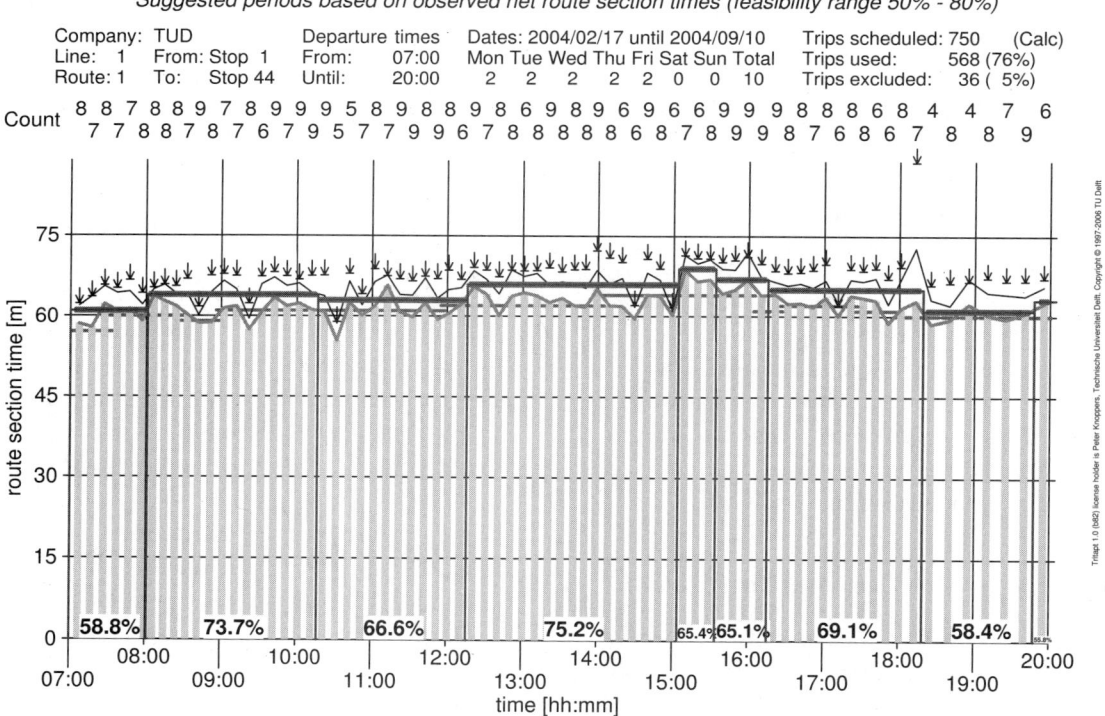

Figure 9. Analysis of automatically suggested running times and periods.

suggested that lies within 2 min of the 85th-percentile observed running time for (almost) every trip in that period.

These algorithms include various rules for expanding, combining, and splitting periods. Other running time analysis programs have similar heuristic algorithms. To the researcher's knowledge, there is no "optimal" formulation for the design of running times and running time periods.

5.1.3 "What-If" Experimentation with Period Boundaries and Allowed Times

This tool allows users to modify both period boundaries and allowed times. The starting point for experimentation can be either the current schedule or the running time periods and allowed times suggested by the program (based on user-selected parameters). Graphical tools allow the user to simply drag period boundaries right or left, split a period, combine periods, and drag proposed allowed times up or down; in response to any change, the program recalculates each period's running time feasibility. Figure 10 shows a user-created set of running time periods and running times and the resulting feasibilities for the same dataset as the previous two figures.

Having a program automatically suggest new periods and allowed times based on user-supplied parameters, while also allowing schedule makers to experiment with and propose their own set of periods and running times, gives schedule makers the power to combine design with analysis. As mentioned earlier, the algorithms that suggest homogeneous periods and running times are "compromisers," not "optimizers." They follow reasonable, systematic rules for determining running times, but given that any solution is an imperfect compromise, users may be able to find solutions they prefer. For example, these algorithms do not consider whether adding a minute of running time might require an extra bus, nor do they consider the burden on passengers of changing the published schedule. Schedule makers can bring this kind of knowledge into the design process; they therefore need the flexibility to modify suggested running times and have the program analyze what will happen.

5.2 Determining Running Time Profiles Using the Passing Moments Method

Once running times for a given route (or route segment) and period of the day are selected, the next step is to divide the chosen route (or segment) time by (smaller) segments, creating a scheduled running time profile (cumulative allowed time from the start of the line). This step must be performed separately for each running time period. For example, take the period 8:06 to 8:42, for which the selected allowed time in Figure 10 was 64 minutes. In the graph shown in Figure 11, the suggested running time profile is shown as the heavy line with

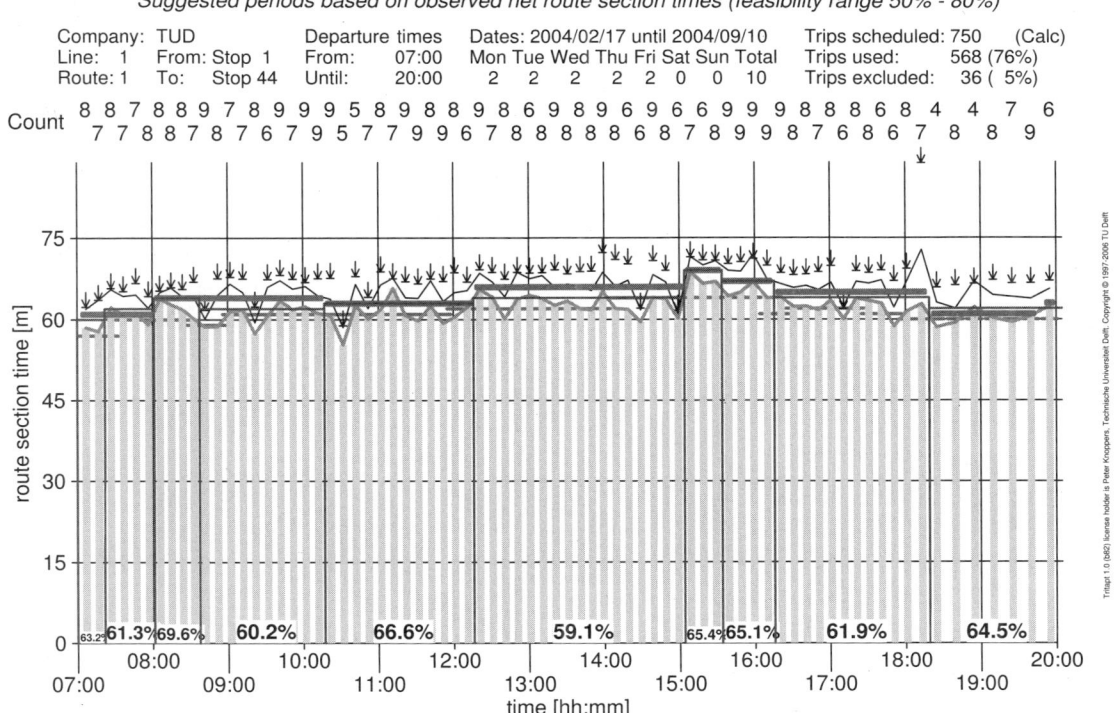

Figure 10. Analysis of user-proposed running times and periods.

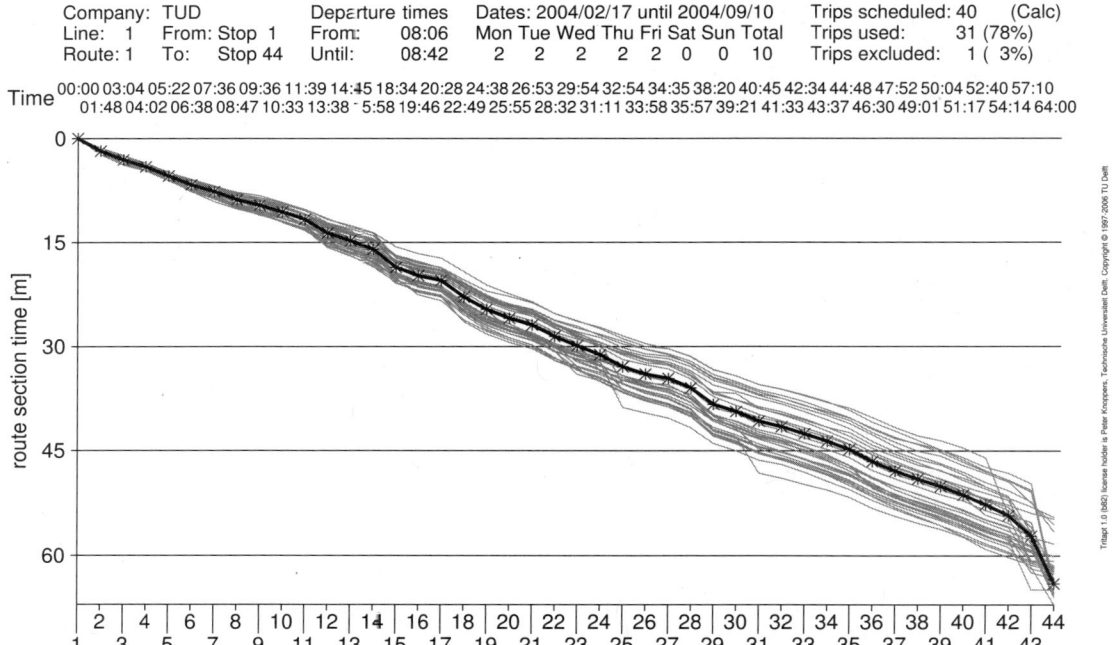

Figure 11. Segment running times or Passing Moment.

asterisks at each stop. To show the relation of the suggested running time profile to observed running time data, this format includes a light line for every observed running time in the selected period, anchored to a start at time 0.

The suggested running time profile uses Muller's Passing Moments method, setting the running time from a timepoint to the end of the line equal to the f-percentile completion time from that timepoint, where f is the feasibility (or attainability) of the overall route time. For example, in Figure 11 the overall route time has 70% feasibility, and so running time from each timepoint to the end of the line is set equal to the 70th-percentile completion time from that timepoint. If running time data is available at the stop level, a data-driven, stop-level running time profile will be created, which can be valuable for passenger information, operational control, and traffic signal priority.

Running time periods and running times accepted in the homogeneous periods analysis are stored in memory and listed in a menu, so that users can choose them one at a time to create running time profiles using the Passing Moments tool.

CHAPTER 6

Tools for Analyzing Waiting Time

Poor service reliability affects both passenger waiting time and crowding. However, traditional methods of analyzing passenger waiting time and crowding, developed in the data-poor age before AVL and APCs, do not account well for the impacts of irregularity upon passenger experience with respect to crowding and waiting time. This chapter presents some methods for analyzing waiting time using AVL data; the next chapter presents methods for analyzing crowding using APC data. These methods have been applied in spreadsheet files, which serve as prototypes of analysis tools that can be applied in AVL-APC data analysis software. (The spreadsheet files are available on the project description web page for TCRP Project H-28 on the TRB website: www.trb.org.)

6.1 A Framework for Analyzing Waiting Time

The researchers developed a new framework for analyzing waiting time, one that accounts for how uncertainty in headway and schedule deviation affects not only how long passengers wait on the platform, but also how much time they have to budget for waiting. That framework is described with mathematical justification in Furth and Miller (8). This section outlines the framework's main features. Then, they are applied to short-headway service in Section 6.2 and to long-headway service in Section 6.3.

6.1.1 Platform Waiting Time

AVL captures data on headways and bus departure times. By making reasonable assumptions about when passengers arrive, and assuming the first bus is not too full for them to board, mean waiting time and the distribution of waiting time can be determined. "Platform waiting time" is the term used for the time passengers actually spend waiting at a stop.

6.1.2 Budgeted and Potential Waiting Time

AVL data can also be used to estimate how long passengers have to budget for waiting. To have a small probability of arriving late at their destination, passengers must plan on waiting longer than the average platform waiting time. While passengers vary in their willingness to accept the risk of arriving late at their destination, a reasonable working assumption is that passengers will accept a 5% risk of arriving late. Therefore, the 95th-percentile waiting time can be interpreted as budgeted waiting time.

Budgeted waiting time can be divided into two parts: the part that passengers actually spend waiting and the remainder, called "potential waiting time." For example, if a passenger budgets 10 min for waiting, but the bus arrives after only 4 min, the 6-min difference is the potential waiting time. Potential waiting time is not spent on the platform; it is spent at the destination end of the trip, where the traveler will arrive 6 min earlier than budgeted. However, because it was set aside for waiting, that time cannot be used as freely as if it had not been so encumbered; therefore, it still represents a cost to passengers. For example, passengers going to work in the morning could not spend their potential waiting time sleeping a few minutes later or staying at home with the kids a few minutes longer. Potential waiting time is a hidden cost associated with waiting, manifested in passengers having to start their trips earlier than they would otherwise have to if waiting time were certain.

6.1.3 Equivalent Waiting Time

Equivalent waiting time is a weighted sum of platform and potential waiting time that expresses passengers' waiting cost in equivalent minutes of platform waiting time. If the weight given to potential waiting time is 0.5, equivalent waiting time is given by

$$W_{equivalent} = W_{platform} + 0.5 * W_{potential}$$

The coefficient 0.5 expresses the cost of a minute of potential waiting time in terms of platform waiting time. Ideally, this parameter value should be estimated based on market research into traveler behavior and preferences. However, 0.5 is a reasonable value consistent with the body of travel demand research (8), lying between 0 (it has a real cost) and 1 (its unit cost should be less than that of platform waiting), and large enough to explain part of the reason that demand models typically assign large relative coefficients to waiting time.

When the unit cost of potential waiting time is 0.5, average equivalent waiting time can also be expressed as the average of mean platform time and budgeted waiting time:

$$W_{equivalent} = 0.5\left(W_{platform} + W_{0.95}\right)$$

where $W_{0.95}$ is the 95th-percentile waiting time.

6.1.4 Service Reliability and Waiting Time

The transit industry has lacked a measure of service reliability that is measured in terms of its impact on customers. Traditional measures of service reliability such as coefficient of variation (cv) of headway and percentage of on-time departures are valid descriptors of operational quality, but they do not express reliability's impact on passengers. For example, how much is it worth to passengers to reduce the headway cv from 0.3 to 0.2, or to improve schedule adherence from 80% to 90%? Because a method of measuring reliability's impact on passengers has been lacking, waiting time has been underestimated, and service reliability undervalued.

Poor service reliability affects passengers mainly by (1) making them wait longer and (2) making them have to budget more time for waiting. (It can also cause crowding, but that is something that can be measured directly.) Equivalent waiting time is a measure that accounts for both of those impacts. Because it includes budgeted waiting time, it is particularly sensitive to service reliability. It is measured in minutes of passenger waiting time, something that can be economically evaluated and compared with, for example, the cost of improving service reliability, or the cost of a headway reduction as an alternative means of reducing passengers' waiting time.

6.1.5 Ideal and Excess Waiting Time

Passenger waiting time can be divided into two parts: ideal and excess. Ideal waiting time is the average waiting time that would result from service exactly following the schedule (35); excess waiting time is the difference between actual and ideal waiting time and is, therefore, the component of waiting time that can be attributed to operational issues. Separating excess from ideal waiting time provides a good idea of the quality of operations and the extent to which passenger service could be improved by improving service reliability. Excess waiting time is also a good measure to use for evaluating contracted service, where the contractor is responsible for operations but not planning; Transport for London uses this measure with its contract bus operators.

Excess waiting time can be a negative value; such a situation could occur, for example, if more service is operated than scheduled.

Until now, the concept of ideal and excess has been applied only to mean waiting time; however, it also applies to budgeted and equivalent waiting time, as the following sections show.

6.2 Short-Headway Waiting Time Analysis

6.2.1 Distribution of Waiting Time

For transit service with short headways, passengers can be assumed to arrive independent of the schedule, effectively in a uniform fashion. If passengers are also assumed to depart with the first vehicle departure after their arrival (i.e., assuming there are no pass-ups), the complete distribution of waiting time can be determined from the set of observed headways. This determination is a step beyond the well-known formula for mean waiting time:

$$E[W] = 0.5 E[H](1 + cv_H^2) \qquad (1)$$

where $E[W]$ and $E[H]$ are mean waiting time and mean headway, respectively, and cv_H is the coefficient of variation of headway, which is the standard deviation of headway divided by the mean. (When applying this formula to a set of observed headways, one should use the "population standard deviation," dividing by n = number of observations, rather than sample standard deviation, which divides by $n - 1$).

Furth and Muller (8) explain how the distribution of passenger waiting time can be derived for both a theoretical headway distribution and for an arbitrary set of observed headways determined from AVL data. The method is best explained by an example. Suppose a route's scheduled headway is 8 min, and six buses were observed with the following headways (in minutes):

Example 1 observed headways : {4,5,7,9,10,13} minutes

With those headways, assuming passengers arrive at random and board the first bus, the waiting time distribution is as shown in Figure 12. At each observed headway, the waiting time distribution steps down in equal steps. Also shown for comparison is the ideal waiting time distribution (i.e., the waiting time distribution that would occur if service had perfectly regular 8-min headways).

In a waiting time distribution, the 95th-percentile waiting time is the value on the horizontal axis that divides the wait-

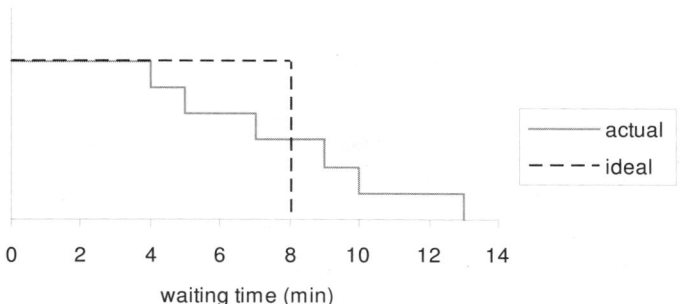

Figure 12. Passenger waiting time distribution for Example 1.

ing time distribution into two parts, with 95% of the area to the left and 5% to the right. For Example 1, the 95th-percentile waiting time equals 10.6 min for the actual headway distribution; for the ideal headway distribution, 95th-percentile waiting equals 7.6 min.

6.2.2 Waiting Time Summary

While the graph of the distribution of waiting time helps explain the relationship between headways and waiting time, it is not a useful format for management reporting or service quality monitoring. Two other formats are therefore offered.

The first format for summarizing passengers' waiting time experience is a summary of platform, budgeted, and equivalent waiting time, as shown in Figure 13. Optionally, the user can show how waiting time breaks out between ideal and excess waiting time. For example, the 10.6 min of budgeted waiting time divides into an ideal budgeted waiting time of 7.6 min and an excess budgeted waiting time of 3.0 min. Likewise, equivalent waiting time, being 7.6 min, divides into ideal and excess parts of 5.8 and 1.8 min, respectively. Therefore, irregularity on this route costs passengers the equivalent of 1.8 min of waiting time.

This format can also be used in a before–after comparison, as shown in Figure 14. Figure 14 is based on a different exam-

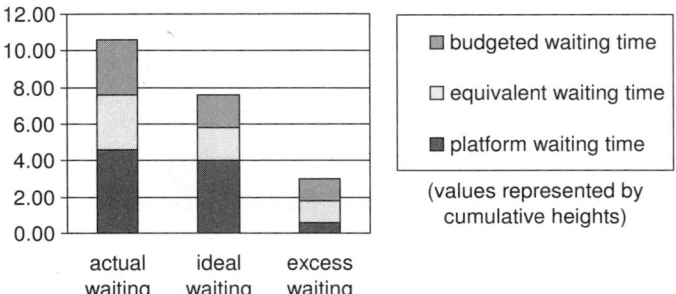

Figure 13. Passenger waiting time summary for Example 1.

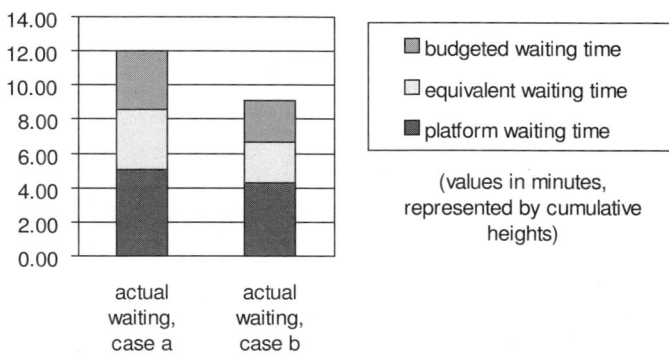

Figure 14. Passenger waiting time comparison for Example 2.

ple, for which scheduled headways in the period of interest are not all equal:

Example 2 scheduled headways:
{5, 8, 8, 8, 8, 8, 8, 8, 9, 9, 9} minutes

With Example 2, two datasets of 100 observed headways each are compared: Case a, with high irregularity (headway $cv = 0.52$), and Case b, with low irregularity (headway $cv = 0.26$). (Actual data values can be found in the spreadsheet file on the project description web page for TCRP Project H-28 on the TRB website: www.trb.org.) The waiting time summary in Figure 14 shows that, with the reduction in irregularity, mean platform waiting falls only a little, from 5.1 to 4.3 min; budgeted waiting time falls much more, from 12.0 to 9.0 min. Combining these waiting components under the composite measure equivalent waiting time shows that the reduction in irregularity saves passengers the equivalent of 1.9 min of waiting time—a result that, if service reliability were not accounted for, would require a headway reduction of almost 4 min, an action that would double operating cost.

This example illustrates how the measures budgeted waiting time and equivalent waiting time reflect the impact of service reliability on passengers.

6.2.3 Percentage of Passengers with Excessive Waiting Times

A second useful reporting format shows the percentage of passengers in various waiting time ranges or "bins." This format can be used to support a service quality standard such as "no more than 5% of passengers should have to wait longer than (scheduled headway + 2) minutes." In its program for certifying bus service quality, the French quality institute AFNOR Certification applies a standard in this format (36). Transit agencies in Paris, Brussels, and Lyon are among those with at least some bus lines certified under this standard.

Calculations for the Example 1 headway data are shown in Table 5. These calculations use a three-bin format, with thresholds of 9 and 11 min, which are 1 min and 3 min beyond the scheduled headway, respectively; the waiting times represented by these bins might be interpreted as "normal," "excessive," and "unacceptable." The last observed headway in the table, which is 13 min long, best illustrates the calculation. Of the passengers arriving during the 13 min, those arriving in the first 2 min of the headway wait 11 min or longer; those arriving in the next 2 min wait 9 to 11 min; and those arriving in the last 9 min wait between 0 and 9 min. Overall, this table shows that 4.2% of passengers wait longer than 11 min.

This format can also offer insights when comparing waiting time distributions. Table 6 compares the percentage of passengers with excessive and unacceptable waiting times for Example 2, Cases a and b. As Table 6 shows, the improved regularity of Case b has dramatically reduced the percentage of passengers with excessive and unacceptable waiting times.

6.3 Long-Headway Waiting Time Analysis

For routes with long headways, most passengers time their arrival at the platform to make a targeted departure. Schedule adherence or schedule deviation is critical to determining passengers' waiting time. Let V equal the schedule deviation of the trip a passenger is trying to meet, defined by

V = departure time − scheduled departure time

Early departures are represented by negative values of V.

As long as schedule deviations are small compared to the scheduled headway, the number of passengers using a given trip will be independent of its schedule deviation or headway. Therefore, unlike with short-headway service, the experience of passengers is the same as the "experience" of buses (i.e., if 15% of the buses were late, it is fair to say that 15% of the passengers had to wait for buses that were running late).

Table 5. Excessive waiting time calculation.

Headway	Waiting Time (Bin Width)			Total
	Normal 0 to 9 min (9 min)	Excessive 9+ to 11 min (2 min)	Unaccept. > 11 min (1000 min)	
4.0	4.0			4.0
5.0	5.0			5.0
7.0	7.0			7.0
9.0	9.0			9.0
10.0	9.0	1.0		10.0
13.0	9.0	2.0	2.0	13.0
Total	43.0	3.0	2.0	48.0
Percentage	89.5%	6.3%	4.2%	100%

Table 6. Passenger waiting time distribution for Example 2.

Case	Waiting Time			Total
	Normal 0 to 9 min	Excessive 9+ to 11 min	Unaccept. > 11 min	
a	84.1%	8.2%	7.7%	100%
b	94.8%	4.4%	0.8%	100%

Therefore, distribution of schedule deviation is a measure of operational performance that relates closely with passenger experience. For interpreting schedule adherence as passenger experience, it is helpful to distribute schedule deviations into more than just the traditional three bins of "early," "on time," and "late." For example, the user might specify thresholds at −1, 0, 3, 5, and 10 min and determine the percentage of trips, and therefore the percentage of passengers experiencing trips, in bins interpreted as unacceptably early, less than a minute early, on time (0 to 3 min), a little late, quite late, and unacceptably late.

As valuable a measure as schedule adherence is, it still does not express an impact on passengers. The researchers were able to extend the waiting time framework to long-headway service in order to determine how much excess waiting time is caused by service unreliability.

6.3.1 Waiting Time Components

Passengers are assumed to have a target arrival time and to arrive at or before this time. The target is set to give passengers a very low probability of missing the bus. Therefore, the target is an extreme value at the lower end of the schedule deviation distribution; the researchers use the 2nd-percentile schedule deviation ($V_{0.02}$), meaning the time by which only 2% of buses will have already departed. Passengers following this policy will miss the bus once every 50 trips, or less (depending on how long before the target they arrive).

Two components of waiting time are not avoidable and are, therefore, part of ideal waiting time, not excess waiting time. Because they depend only on the planned headway, they are not a subject of AVL data analysis; nevertheless, for completeness, they are mentioned here:

- **Schedule inconvenience:** a well-known form of hidden waiting time that arises from departures or arrivals not being scheduled when passengers want to travel. It is manifested in passengers who are going to work arriving before their work start time because the next bus would get them there too late. Likewise, after work, passengers may consistently have time to kill between leaving work and the next scheduled departure.
- **Synchronization time:** the cost, in equivalent minutes of waiting, of ensuring that one is at the station by the target arrival time. To be sure of hitting that target, many passen-

gers will arrive early because of uncertainty in access time, the limits of human punctuality, and risk aversion. The waiting time between when passengers arrive and the ideal arrival time is part of synchronization time.

Schedule inconvenience and synchronization time are unavoidable; they can be "blamed" on planning, as they are the inevitable consequence of designing a service with a long headway. In contrast, random waiting time that occurs after the ideal arrival time is due to service unreliability and is, therefore, excess waiting time. Its components, and their relation to schedule deviations, are shown in Figure 15. Excess platform waiting runs from the passengers' target arrival time to the bus's actual departure time. Excess budgeted waiting time runs from the target arrival time to the 95th-percentile schedule deviation ($V_{0.95}$). As in the case of short-headway waiting, passengers cannot plan on being picked up at the average departure time; they have to budget for an extreme value. The researchers assume they budget for the 95th-percentile schedule deviation; with that value, passengers will arrive late 5% of the time.

Excess budgeted waiting time, being the spread in the schedule deviation distribution between $V_{0.02}$ and $V_{0.95}$, can be thought of as the time between the earliest likely and latest likely departure times. The greater the service unreliability, the greater the spread, and the more time passengers have to budget for waiting. Of course, not all of budgeted waiting time is actually spent waiting on any given day; part is spent waiting, and the remainder is potential waiting time, just as with short-headway service. Like schedule inconvenience, potential waiting time is manifested in passengers arriving earlier than planned at their destination. The difference is that potential waiting time varies from day to day and cannot be counted on in planning daily activities, while schedule adherence is not random, and can be counted on, making it less of a burden to passengers.

Using the notation that $E[V]$ equals mean schedule deviation, and using the weight of 0.5 suggested earlier for combining potential waiting with platform waiting, summary measures of excess waiting time can be calculated as follows:

$$\text{mean excess platform waiting} = E[V] - V_{0.02}$$
$$\text{excess budgeted waiting} = V_{0.95} - V_{0.02}$$
$$\text{mean potential waiting} = V_{0.95} - E[V]$$
$$\text{equivalent excess waiting} = (\text{excess platform waiting})$$
$$+ 0.5 * (\text{potential waiting})$$
$$= 0.5\,(\text{excess platform waiting}$$
$$+ \text{excess budgeted waiting})$$

6.3.2 Example Analyses

Example schedule deviation and waiting time summary reports for long-headway service are shown in Figure 16. These figures compare two example cases, each based on 100 synthesized observations:

- Case No-OC is a route with relatively poor reliability. Schedule deviation has a standard deviation of 2.2 min, so that only 72% of departures are in an on-time window of 0 to 5 min late.

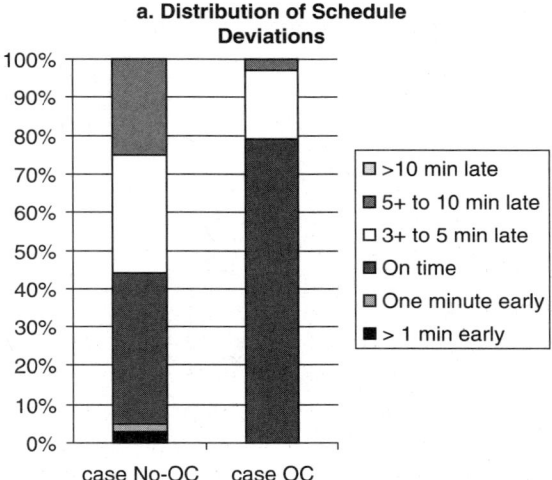

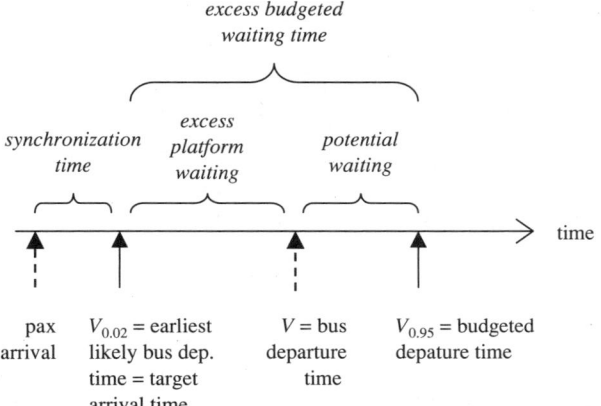

Figure 15. Long-headway waiting time components.

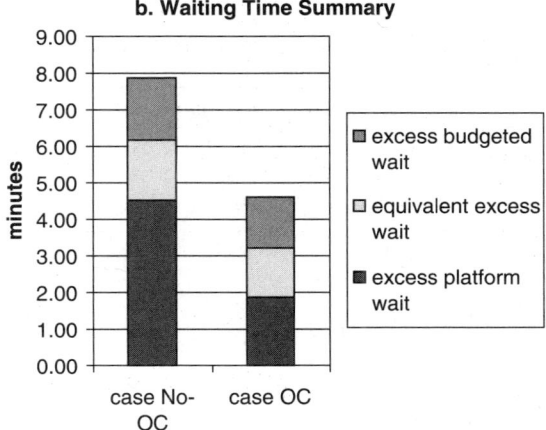

Figure 16. Long-headway schedule deviation and waiting time analysis.

Table 7. Waiting time component comparison.

Component	Case No-OC (min)	Case OC (min)	Change (min)
2nd-percentile schedule deviation	-1.0	0.1	1.1
Mean schedule deviation	3.48	1.88	-1.60
95th-percentile schedule deviation	6.8	4.6	-2.2
Mean excess platform wait	4.5	1.8	-2.7
Equivalent excess wait	6.2	3.2	-3.0
Excess budgeted wait	7.9	4.6	-3.3
Potential wait	3.4	2.8	-0.6

- Case OC is the same route, and has the same underlying variability, as Case No-OC, but operational control is applied by holding, and the scheduled departure time has been shifted earlier by 2.2 min. With that schedule adjustment, the earliest 30% of departures will be held; as a result, they are assumed to have random schedule deviations between 0 and 1.5 min.

Figure 16(a) shows a strong improvement in operational quality reflected in the distribution of schedule deviations. The 5% of trips that were early disappear, and the percentage of trips in the 0- to 5-min window rises from 70% to 97%. The percentage of trips in the highest quality category (0 to 3 min late) rises from 39% to 79%.

The waiting time summary found in Figure 16(b) expresses these results in terms of impact to passengers. Mean platform waiting time falls by 2.7 min because passengers do not have to arrive before the scheduled departure time and the average schedule deviation has been reduced considerably. Counting changes in potential waiting time, there is net reduction in equivalent waiting time of 3 min.

Table 7 provides additional detail on how operational control shrinks the spread between early and late schedule deviations. The 2nd-percentile schedule deviation rises by 1.1 min and the 95th-percentile schedule deviation falls by 2.2 min, shrinking excess budgeted waiting time by 3.3 min.

Because potential waiting time has most to do with the upper end of the schedule deviation distribution, actions that reduce the upper end (gross lateness) tend to most affect potential waiting time, while actions that reduce the lower end (earliness) tend to reduce platform time, as in the example. Because platform waiting time affects passengers more strongly than potential, actions that reduce earliness are therefore particularly effective.

The mean waiting time calculations do not account for the (small) percentage of passengers who miss their bus (perhaps because the bus was early) and have to wait a full headway for the next one. The reason this percentage is not taken into account is two-fold. First, the expected penalty for missing the bus is $0.02\,h$, where h is the scheduled headway, and 0.02 is the probability of missing the bus. Because this quantity depends on the planned headway rather than on a schedule deviation, it does not contribute to excess waiting time because it will be the same regardless of service reliability. (That would not be the case if the target arrival time were not percentile-based, e.g., if it were set at 1 min before the scheduled departure time, for example.)

Second, while the immediate impact of an early bus is to make a few passengers wait a long time, in the long term the impact of early buses is to make all passengers arrive earlier at the bus stop every day. For example, a passenger who misses a bus on a route with a 20-min headway suffers a 20-min waiting time penalty that day. However, for maybe the next 100 days, the passenger will arrive at the stop 2 min earlier to be sure to not miss the bus again, which costs him 200 minutes of additional waiting. As this simple example shows, the impact of earliness on waiting time is accounted for by setting the target arrival time to the 2nd-percentile schedule deviation, which is sensitive to earliness.

To reiterate, determining extreme values from data requires large sample sizes. A rule of thumb is that there should be at least 5 observations outside the extreme value estimated. Therefore, about 250 observations are needed to make a reliable estimate of the 2nd-percentile schedule deviation. AVL will provide that kind of sample size, but it may be necessary to aggregate over a considerable date range and/or over a period of the day with several scheduled trips.

CHAPTER 7

Tools for Analyzing Crowding

Crowding is important to passengers for their comfort; to operations, because it can slow the boarding and alighting process; and to planning, as a measure of efficiency. These different viewpoints need different measures derived from passenger count data.

The planning viewpoint is concerned with average load at the peak volume point, that is, the segment whose average load is the greatest. Schedule planning often uses peak point load to determine headway, using a nominal design capacity. Because this measure is a single number that is widely understood, it is not covered further. However, impacts on operations and on passengers are strongly affected by the random distribution of passenger crowding, something that can only be analyzed well with the large samples that APCs can afford.

This chapter describes methods for analyzing passenger crowding. These methods have been programmed as prototypes in a spreadsheet file which is available on the project description web page for TCRP Project H-28 on the TRB website: www.trb.org.

7.1 Distribution of Crowding by Bus Trip

For both the passenger and operations viewpoints, load can be examined on every segment of a trip. However, for most purposes, analysts want to focus on the most crowded segment (the maximum load segment) of each trip.

The maximum load segment of any trip may differ from the route's peak volume point. If averaged over many trips, the average maximum load will often be greater, and cannot be smaller, than average load at the peak volume point.

Average maximum load is a measure suggested by the *Transit Capacity and Quality of Service Manual* (*TCQSM*) to characterize level of service with respect to crowding (27). The *TCQSM* defines six levels of service (LOSs) (A through F) and suggests thresholds based on the number of seats and amount of available standing space per standee. The examples in this section use the thresholds shown in Table 8; they were determined using *TCQSM* default values and assuming a 40-ft bus with 36 transverse seats, 6 longitudinal seats, a stairwell for the rear door, and 6 ft of unused length at the front of the bus. (For greater detail, the *TCQSM*'s LOS F has been subdivided into levels F1 and F2.)

However, while these thresholds account well for passenger comfort levels on an individual trip, they do not mean much when applied to an "average trip." Neither operations nor passengers care much about *average* crowding. What really matters is the distribution of crowding and, in particular, extreme values. Because of the large sample sizes afforded by APC data, analysts can derive, from maximum load observations, distributions of both trips and passengers by crowding level.

To illustrate, the researchers analyzed 30 observations of peak-hour trips on a certain route and found that the mean value of maximum load was 40.3. (The data and analysis described in this chapter can be found in the spreadsheet file on the project description web page for TCRP Project H-28 on the TRB website: www.trb.org.) With the example 42-seat buses, the *TCQSM* would rate this route-period in LOS C. On average, load is less than the number of seats, which deceptively suggests that everybody should get a seat and that there should be little problem of crowding interfering with boarding and alighting. In fact, the distribution of maximum load over those 30 trips, shown in Figure 17, offers quite a different picture. About 47% of the trips had standees (load > 42); and 20% of the trips were either "crowded" or "overcrowded" (load > 62), which could seriously affect running time. Yet, 27% of the observed trips had at least half their bus seats empty, suggesting a possible bunching problem.

7.2 Distribution of Crowding Experience by Passenger

7.2.1 Classification of Crowding Experience

Measures of passenger service quality should use passengers, not bus trips, as units and should adopt the passenger's viewpoint. Crowding experience from the passenger's viewpoint can be classified as follows:

CHAPTER 8

Passenger Count Processing and Accuracy

Accuracy of automated passenger counts may be reduced by many types of errors, including counting error, location error, attribution error (i.e., attributing counts to the wrong trip), modeling error (e.g., assumptions about loops), and sampling error. It is also important to distinguish between the accuracy of raw counts and that of screened and corrected counts, and between the accuracy of directly measured items (ons and offs by stop) and aggregate measures such as load, passenger-miles, and trip-level boardings.

Finally, for any type of error, it is important to distinguish between bias (systematic error) and random error. While random error, like sampling error, shrinks with increased sample size, correcting for bias is usually impractical; therefore, controlling bias becomes far more important than controlling random error.

To a large extent, this chapter and the following one repeat material originally published in Furth et al. (7).

8.1 Raw Count Accuracy

The accuracy of raw counts tends to be the focus of vendors and many buyers. Kimpel et al. recently studied Tri-Met counts and found statistically significant bias for one of the two bus types tested (bus type affects how sensors are mounted)—an average overcount of 4.24% for ons and 5.37% for offs (37). Overall, standard deviation of random count error was found to be rather large—about 0.5 passenger per stop for both ons and offs, for a coefficient of variation (cv) of 0.37. This value is surprisingly large; the researchers suspect newer systems are more precise.

Kimpel et al. suggest applying correction factors to overcome biases. However, few agencies can afford the research needed to establish the level of systematic over- or undercount. They need counts whose biases are small enough to live with. Less onerous are bias corrections established by the APC vendor. One vendor includes in its processing software correction factors for counts of 1, 2, and 3+ passengers, yielding non-integer corrected counts.

Test criteria for APC equipment often fail to distinguish between random and systematic error. For example, the criterion "the count should be correct at 97% of stops" does not consider whether there might be a tendency to over- or undercount. Another weakness of this criterion is that many stops may have zero ons and offs, which is rather easy to count correctly.

To control bias, tests should require that the ratio of total counted ons to total "true" ons be close to 1. Using Tri-Met's random stop-level error cv of 0.37, the hypothesis that the on counts have no systematic error can be accepted at the 95% confidence level if this ratio is in the range $1 \pm 0.72\sqrt{n}$, where n equals the number of stops contributing to the test total. A less stringent test would allow a small degree of bias, for example, 2% (partly in recognition that the "true" count may itself contain errors); then the acceptance range becomes $1 \pm (0.02 + 0.72\sqrt{n})$, which, with n equal to 5000, is the range 1 ± 0.03.

One of STM's tests is that, at the trip level, the average absolute deviation between automated and manual counts of boardings should be less than 5% of average trip boardings. Because it uses absolute deviations, this test masks systematic error. However, the strict criterion of 5% effectively forces both random and systematic error to be small.

Acceptance tests should specify the screening criteria and the maximum percentage of trips (or blocks of trips) rejected, and then apply accuracy criteria to the remaining data. STM provides a good example: it rejects trips with an imbalance of 5 or more passengers, requires that no more than 85% of trips be rejected, and applies accuracy criteria to the remaining trips.

8.1.1 Measuring Ground Truth

One problem in testing APC accuracy is the difficulty of observing ground truth. Conventional manual counts can have greater counting error than a good APC. One vendor insists that clients use video cameras, at least one per door, in acceptance tests, as the vendor does in research and development. When the CTA tested its new APCs, it enhanced the reliability of its manual counts by using a three-person crew staffed by managers and data analysts who had a stake in the outcome and, therefore, reason to be meticulous in counting accurately.

8.1.2 Block-Level Screening

Because APCs count both ons and offs, large errors (e.g., due to malfunctioning equipment) are easy to spot based on a large difference between total on and off counts over a bus's daily duty ("block"). Screening criteria vary. Tri-Met rejects blocks whose on and off totals differ by more than 10%; King County Metro requires that a block's total offs be no more than 7% below or 15% above the block's on total. (The asymmetric criterion is a concession to recognized counting bias with its older APC system.)

8.1.3 Accuracy of Load and Passenger-Miles Measurements

Because the accuracy of load and passenger-miles measurements depends not only on raw count accuracy, but also on the processing system's ability to parse blocks into trips and deal with on-off imbalances, accuracy of load and related measures is a good system test and deserves to be examined in its own right. STM sets a good example, requiring that the average absolute error in departing load be less than 5%.

Relative errors in load can be much greater than those of on and off counts. For example, Kimpel et al. found that, while systematic error for Tri-Met's on and off counts were below 2%, it was 6% for departing load (37). Because passenger-miles are a weighted accumulation of loads, one can expect its bias to be similar to that of load.

One cause of load errors can be the balancing method used. Kimpel et al. determined load using their own trip-level balancing method, because the block-level balancing algorithm used by Tri-Met yielded loads with greater bias. Block-level balancing biases loads upward because upward errors are permitted to propagate through the day, while propagation of downward errors is limited because of a restriction against negative departing loads.

8.2 Trip-Level Parsing

Screening based on on-off balance protects on and off totals from substantial errors. However, because of a phenomenon called drift, substantial errors can still develop in calculated load (accumulated ons minus accumulated offs) and passenger-miles (a weighted sum of segment loads), even with small errors in raw counts. To illustrate, suppose each trip has 2 excess ons. After five trips, load on every segment will appear to be 10 passengers greater than it actually is.

APC processing software controls drift by regularly resetting the system state (i.e., load on the bus) at points of known load. On most transit routes, these points are terminals or layover points at which, either by custom or operating rule, the through-passenger load is zero. Blocks are then parsed at known-load points into sections that may be called "sub-blocks," which are usually single trips or round trips, with load at the start and end of each sub-block set to zero.

Data structures and processing software need to account for some trip ends being points of zero load and others not; moreover, there may be points of known zero load that are not trip ends.

8.2.1 End-of-Line Identification and Activity Attribution

The earlier review of automatic location measurement pointed out difficulties often encountered in correctly identifying the end of the line. This problem is most severe in older APC systems that lack sign-in data and suffer from low schedule-matching rates.

When the general location of a route endpoint can be correctly matched, the end-of-line arrival and departure time issues that vex running time analysis are not a concern for passenger counts; rather, the main challenge is attributing ons and offs to the right trip. At a simple route terminal, the usual procedure is to attribute alighting passengers to the arriving trip and boarding passengers to the departing trip. Sometimes, the boarding and alighting activity are sufficiently simultaneous that APCs make a single stop record, which processing software has to split. Sometimes many stop records will be generated at the terminal (and at other stops as well) as the bus may go through several cycles of opening doors to let passengers board, then closing doors to preserve or keep out the heat. On-board APC analyzers vary in their ability to take external inputs (e.g., odometer pulses) and use them in determining when to close a stop event and start a new one. Off-line processing software has to have the flexibility to recognize and handle both single- and multiple-record cases.

8.2.2 Inherited and Bequeathed Passengers

Operating practices for some routes allow passengers to remain on board at the end of one trip in order to ride on the next trip. One example is a route that ends with a loop; another example is a pair of interlined routes for which nominally transferring passengers actually remain on board. Data structures have to identify which route ends are not necessarily zero-load points and recognize passengers inherited from a previous trip.

To permit trip-level data analysis, the most direct way to deal with inherited passengers is for databases to include an extra stop record at the start and end of each trip indicating the number of passengers inherited from the previous trip and bequeathed (left on board) to the next trip. Manual ride check forms often start and end with a row for passengers left on board.

An alternative arrangement, proposed by at least one vendor, makes no record of inherited passengers but assumes that any imbalance in ons and offs at the trip level can be explained as inherited or bequeathed passengers. If a trip has more offs than ons, the difference is treated as passengers inherited at the start of the trip; if more ons than offs, the difference is treated as bequeathed passengers at the end of the trip. However, this shortcut has several shortcomings. Imbalance can be due to counting errors as well as inherited passengers. In the face of counting errors, this approach will violate the law of conservation, because it does not guarantee that the number of passenger bequeathed by one trip equals the number inherited by the next. By forcing corrections to be positive (i.e., as opposed to correcting imbalance by reducing ons or offs), it biases upward the number of passengers; and by concentrating the corrections at the route ends, it increases average passenger trip length. Together, these factors combine to bias passenger-miles upward. Also, this approach cannot resolve imbalances on routes with loops at both ends.

8.2.3 Routes Ending in Loops

Many transit routes end in loops that lack a natural terminal point at which buses always empty out. Examples are radial routes with a loop at the suburban end for wider coverage and commuter routes with a collection/distribution loop through the downtown, such as NJ Transit routes into Philadelphia. There are three ways to deal with attributing passengers who board or alight on these loops; each approach has important implications for data structures.

The Round-Trip Approach

The simplest approach is treating the round trip as a single trip. However, the way an agency defines its trips is part of a business model that carries into its scheduling database, with which the APC database must be consistent. Therefore, this solution is only available to the extent that the schedule database is constructed in terms of round trips.

The Terminal-Stop Approach

A second way—perhaps the most common—to deal with a loop is to designate a terminal stop somewhere in the loop; there may be a short layover scheduled there. Through load at this point is treated as passengers inherited by the trip leaving the loop.

The Overlapping Loop Model for Short Loops

A third approach, used by NJ Transit on its Philadelphia routes (*38*), is to model a bus in a loop as serving two trips at once, attributing alightings to the trip entering the loop and boardings to the trip exiting the loop. Figure 19 illustrates the overlapping loop model. With this model, boardings and alightings occurring within a loop are attributed to the trip that "naturally" serves those passengers; there are no explicit inherited passengers, avoiding messy apparent transfers. This model is also well suited to making balancing corrections.

The overlapping loop model relies on two assumptions about passenger travel:

- **General Loop Assumption:** No passenger rides around the entire loop.
- **Short Loop Assumption:** No passenger's trip lies entirely within the loop.

Discretion is needed in defining the boundary between the entering and exiting trip for boardings. For NJ Transit routes into Philadelphia, boardings on the exiting trip clearly begin at the first Philadelphia stop. However, on some routes with loops, the exiting trip's boardings may begin at the last trunk stop (Stop A in Figure 19) if travel time around the loop is smaller than the service headway, in which case passengers waiting at A' have nothing to lose by boarding at A and circling the loop.

With the overlapping route model, a short loop effectively serves as a fixed-load point for parsing, screening, and on-off balancing. For the entering trip, load is fixed at zero when it has finished serving the loop; for the exiting trip, load is fixed at zero as it begins serving the loop.

An example of fixing and balancing load on a loop with four stops (A through D) is shown in Table 9. In Table 9(a), there is an imbalance of 2 excess offs; however, after the data is split into entering and exiting trips, one can see that the entering trip actually has 4 excess offs and the exiting trip 2 excess ons. Corrections are made in Table 9(b) to the entering and exiting trip separately, following the balancing procedure described later in Section 8.3.

Applying the overlapping route model requires data structures that recognize loop start and end points and the relationship between the trips entering and exiting the loop. The model can be used only as part of processing the raw counts; corrected counts are returned to a database without overlapping routes. If the model is used in such a manner, a stop in the loop is designated as the terminal where stop records are inserted that give the number of bequeathed and inherited passengers. It is easy to show that

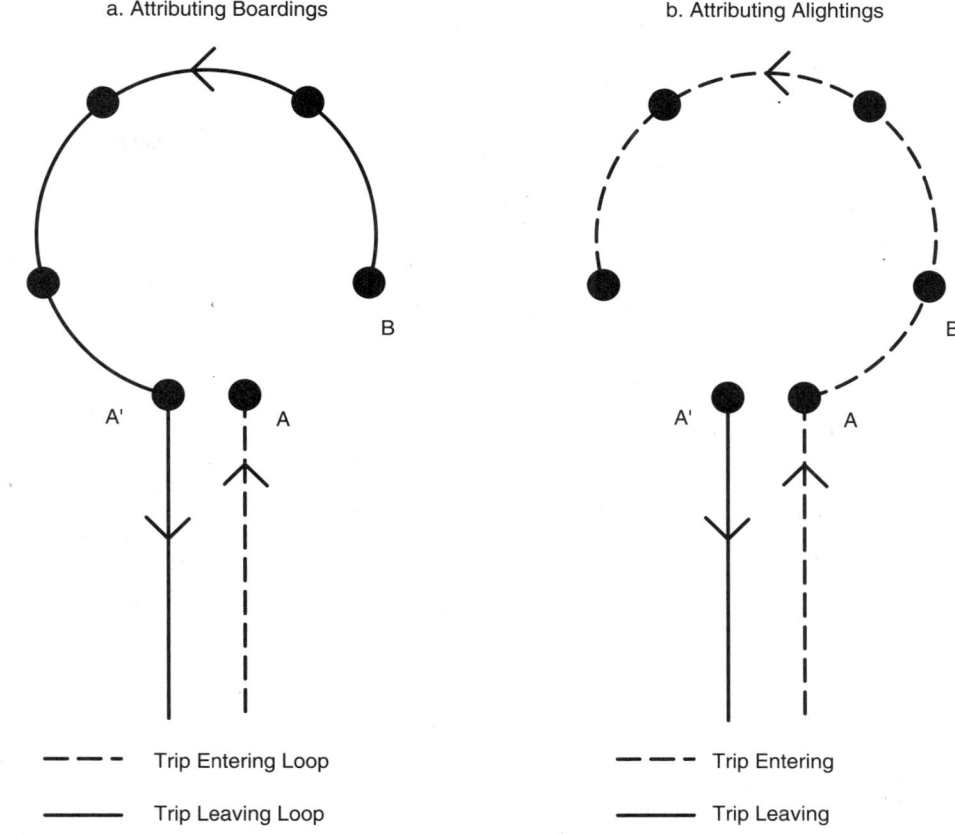

Figure 19. Overlapping loop model.

Inherited Passengers = EarlyOns + LateOffs (2)

where *EarlyOns* equal ons in the loop before the terminal, and *LateOffs* equal offs in the loop after the terminal. As shown in Table 9(c), inherited passengers = (5+5) + 9 = 19.

It is also possible to have an overlapping route data structure in the final database of corrected counts. Such a data structure would allow analyses that treat the entire loop as part of both the trip entering and the trip exiting the loop, with passenger movements on the loop appropriately attributed. That data structure requires methods to properly deal with overlapped sections. For example, the actual vehicle load within the loop is equal to the sum of load on the entering trip and load on the exiting trip. As another example, care must be taken not to double count operating statistics on the loop such as vehicle-miles or schedule deviations.

8.2.4 Routes Without a Fixed-Load Point or Short Loop

The few routes that do not have a zero-load point or short loop at either end pose a challenge for preventing drift. Examples are downtown circulators and routes with long loops at both ends. Simply letting ons and offs accumulate all day long, and taking the difference as load, invites drift errors if ons are overcounted early in the day, even after total ons and offs over the day have been balanced.

The only solution that has been suggested involves operator intervention: having the operator count and enter, via the control head, the load at a certain point in each cycle, preferably a point where the load is normally low. While requiring operator input violates the traditional design philosophy of APCs, it may be the only way of ensuring reliable load data on routes without zero-load points. Buyers in this situation can request that a new APC system support an operator-initiated "observed load event," including an automatic prompt at the designated location.

8.2.5 Accounting for Operator Movements

Most operator off/on movements occur at layover points when the bus is empty. Absent counting errors, an operator who gets off and back on an empty bus at a layover point can be detected because such movements cause an apparent through load (arriving load minus offs) of −1, or a still more negative number if the operator gets off and on several times, for example, to adjust a mirror.

Table 9. Balancing load on a loop.

(a) Splitting Data into Entering and Exiting Trips

Stop	Combined			Entering Trip			Exiting Trip		
	Off	On	Load	Off	On	Load	Off	On	Load
Inherited			0			0			0
Before loop	6	42	36	6	42	36			0
A	10	5	31	10		26		5	5
B	10	5	26	10		16		5	10
C	10	5	21	10		6		5	15
D	10	5	16	10		-4		5	20
After loop	18	0	-2				18	0	2
Total	64	62		46	42		18	20	

(b) Correcting Entering and Exiting Trips

Stop	Entering Trip			Exiting Trip		
	Off	On	Load	Off	On	Load
Inherited			0			0
Before loop	6	44	38			0
A	10		28		5	5
B	9		19		5	10
C	10		9		4	14
D	9		0		5	19
After loop				19	0	0
Total	44	44		19	19	

(c) Splitting Trips at Terminal (Stop) C

Stop	Entering Trip			Exiting Trip		
	Off	On	Load	Off	On	Load
Inherited			0			
Before loop	6	44	38			
A	10	5	33			
B	9	5	29			
C (off only)	10		19			
Bequeathed / inherited			19			19
C (on only)					4	23
D				9	5	19
After loop				19	0	0
Total	35	54		28	9	

The researchers are not aware of efforts in APC data processing to identify and exclude operator movements; clearly, there is a need for such algorithms, especially those that work well in the presence of possible counting error.

8.3 Trip-Level Balancing Methods

Comparing on and off totals gives APCs a built-in error check. Large errors can simply be screened out; that is part of the advantage of the large sample size that comes with automated data collection. Counts with smaller imbalances will be accepted, leaving the question of how counts should be corrected in order to balance ons and offs.

Besides achieving on-off balance, corrections aim to prevent negative loads. The algorithms the researchers have seen consider only departing load. A more stringent feasibility test looks also at through load. For example, suppose a bus arrives at a stop with a load of 2 passengers; on-off counts then indicate that 6 passengers got off and 5 got on. The departing load is calculated to be 1, a positive number that raises no alarm. However, something about these figures is not right; how could 6 passengers get off when only 2 were on the bus? This discrepancy is clear when one calculates that through load based on these figures is −4. Negative through loads can occur if passengers (or the bus operator) get on and off at the same stops (e.g., passenger steps on the bus, finds out it is the wrong bus, and steps off); but that is a rather rare occurrence, except perhaps at terminals. To account for an occasional balker or an operator getting off and back on, correction algorithms can set −1 as the lower limit on through load, something illustrated in the following example. Offs occurring at terminals well after the bus has arrived and discharged its load are likely to be balkers, and processing algorithms may explicitly seek to identify and deal with them.

8.3.1 Selecting Target On/Off Total

In a sub-block, boundary conditions require that the difference between total ons and offs equal the difference between bequeathed and inherited passengers. If the totals do not balance, should total ons be adjusted to match total offs, or vice versa, or should the correction be shared between the two?

If the counting system has known error patterns, they can be taken into account in selecting the target. Let

T^*_{on}, T^*_{off} = target (corrected) sub-block ons and offs total
T_{on}, T_{off} = block level raw totals for ons and offs
M = bequeathed−inherited passengers for the sub-block (given)
k_{on}, k_{off} = external adjustment factors for ons and offs
s^2_{on}, s^2_{off} = variance of sub-block errors for on and off totals
$c_{on} = 1/s^2_{on}, c_{off} = 1/s^2_{off}$

The external adjustment factors are correction factors an agency might have that account for known systematic bias; for example, $k_{on} = 1.03$ implies that ons have a systematic undercount of 3%.

From information theory, the best (least variance) estimates are given by

$$T^*_{on} = \frac{c_{on} k_{on} T_{on} + c_{off} k_{off} T_{off} + M c_{off}}{c_{on} + c_{off}} \quad (3)$$

$$T^*_{off} = T^*_{on} - M \quad (4)$$

When $M = 0$, this expression is a weighted average of adjusted on and off totals, using relative certainty as weights. For example, if $c_{on}/c_{off} = 3$ and a sub-block had 4 excess ons, the correction would be to reduce ons by 1 and increase offs by 3. Even if on counts are known to be more accurate than off counts, or vice versa, the target is more accurately estimated using information from both sets of counts, rather than ignoring the weaker set.

8.3.2 Distributing Corrections

One simple way to distribute corrections that has already been mentioned is to put them at the end (or start) of the sub-block. The method used in TriTAPT is to make the corrections to the largest counts, the assumption being that counting errors are more likely to have occurred there.

Another systematic method uses proportional corrections. One simply multiplies all the ons in a sub-block by the correction factor $f = T^*_{on}/T_{on}$, and all offs by the correction factor $f = T^*_{off}/T_{off}$. This method has two desirable properties: first, it leaves the alighting and boarding centroids unchanged, leaving average passenger trip length unaffected; second, it calls for greater corrections where counts were greater, which is consistent with the notion that the bigger the count, the more likely an error.

A rounding procedure is illustrated in the example that follows. To round ons, first generate the profile of cumulative adjusted ons; round the cumulative profile; and then generate rounded ons by stop as the difference between successive cumulative ons. As mentioned earlier, this rounding procedure can be applied to force counts in the database to be integers, or it can be applied when generating reports.

8.3.3 Correcting Negative Loads

After a sub-block is balanced, calculated loads may be negative at one or more points along a route. Most commonly checked is departing load; however, a stronger test is for through load (departing load minus ons). There should be thresholds for negative departing load and negative through load beyond which a trip should be rejected.

Trips not rejected should be adjusted to eliminate negative loads. One option, illustrated in this section, is to allow through loads of −1 to account for an operator exiting and re-entering an otherwise empty bus, or a passenger boarding and then alighting an empty bus (e.g., upon discovering it was the wrong bus).

If there are multiple points with negative load, the point of greatest violation should be corrected first, because its correction is likely to cure negative loads elsewhere. The point of correction becomes a new sub-block boundary, dividing the sub-block into two new sub-blocks which are then balanced, making the procedure recursive.

A balancing example is given in Table 10. Original counts [Table 10(a) and (b)] have 36 ons and 34 offs. First, target on and off totals are calculated to be 35, then ons and offs are adjusted and rounded.

A check for negative loads in Table 10(c) finds the greatest violation at Stop 5. Through load there is −4, but because through loads of −1 are allowed, the violation is −3. The trip is split at Stop 5 into two new sub-blocks, which are balanced in Table 10(d) through (k). Stop 5's offs belong to the early sub-block (the one ending at Stop 5), and its ons to the late sub-block.

The early sub-block [Table 10(d) and (e)] begins with an imbalance of −3 (25 ons, 29 offs, and a target difference of −1).

Table 10. Balancing initial load and correcting negative load.

(a) Balance Ons

Stop	Input Ons	Cumulative Ons	Scaled Cumulative	Rounded Cumulative	Balanced Ons
1	12	12	11.67	12	12
2	8	20	19.44	19	7
3	6	26	25.28	25	6
4	0	26	25.28	25	0
5	2	28	27.22	27	2
6	5	33	32.08	32	5
7	2	35	34.03	34	2
8	0	35	34.03	34	0
9	1	36	35.00	35	1
10		36	35.00	35	0
Total	36				35
Target	35		$f = 0.972$		

(b) Balance Offs

Stop	Input Offs	Cumulative Offs	Scaled Cumulative	Rounded Cumulative	Balanced Offs
1		0	0.00	0	0
2	2	2	2.06	2	2
3	4	6	6.18	6	4
4	10	16	16.47	16	10
5	12	28	28.82	29	13
6	0	28	28.82	29	0
7	1	29	29.85	30	1
8	0	29	29.85	30	0
9	3	32	32.94	33	3
10	2	34	35.00	35	2
Total	34				35
Target	35		$f = 1.029$		

(c) Check for negative load

Stop	Off	On	Thru Load	Dep Load	Violation	Comment
1	0	12	0	12		
2	2	7	10	17		
3	4	6	13	19		
4	10	0	9	9		
5	13	2	-4	-2	-3	split here
6	0	5	-2	3	-1	
7	1	2	2	4		
8	0	0	4	4		
9	3	1	1	2		
10	2	0	0	0		

(continued on next page)

Table 10. (Continued).

(d) Balance Ons, Early Subblock

Stop	Input Ons	Cum Ons	Scaled Cum	Round'd Cum	Bal'c'd Ons
1	12	12	12.96	13	13
2	7	19	20.52	21	8
3	6	25	27.00	27	6
4	0	25	27.00	27	0
5	0	25	27.00	27	0
Total	25				27
Target	27		$f=$ 1.080		

(e) Balance Offs, Early Subblock

Stop	Input Offs	Cum Offs	Scaled Cum	Round'd Cum	Bal'c'd Offs
1	0	0	0.00	0	0
2	2	2	1.93	2	2
3	4	6	5.79	6	4
4	10	16	15.45	15	9
5	13	29	28.00	28	13
Total	29				28
Target	28		$f=$ 0.966		

(f) Check for Neg Load, Early Subblock

Stop	Offs	Ons	Thru Load	Dep Load	Violation
1	0	13	0	13	
2	2	8	11	19	
3	4	6	15	21	
4	9	0	12	12	
5	13	0	-1		

(g) Balance Ons, Late Subblock

Stop	Input Ons	Cum Ons	Scaled Cum	Round'd Cum	Bal'c'd Ons
5	2	2	1.60	2	2
6	5	7	5.60	6	4
7	2	9	7.20	7	1
8	0	9	7.20	7	0
9	1	10	8.00	8	1
10	0	10	8.00	8	0
Total	10				8
Target	8		$f=$ 0.800		

(h) Balance Offs, Late Subblock

Stop	Input Offs	Cum Offs	Scaled Cum	Round'd Cum	Bal'c'd Offs
5	0	0	0.00	0	0
6	0	0	0.00	0	0
7	1	1	1.17	1	1
8	0	1	1.17	1	0
9	3	4	4.67	5	4
10	2	6	7.00	7	2
Total	6				7
Target	7		$f=$ 1.167		

(i) Check for Neg Load, Late Subblock

Stop	Offs	Ons	Thru Load	Dep Load	Violation
5	0	2	-1	1	
6	0	4	1	5	
7	1	1	4	5	
8	0	0	5	5	
9	4	1	1	2	
10	2	0	0	0	

(j) Balance Counts

Stop	Offs	Ons	Thru Load	Dep Load
1		13	0	13
2	2	8	11	19
3	4	6	15	21
4	9	0	12	12
5	13	2	-1	1
6	0	4	1	5
7	1	1	4	5
8	0	0	5	5
9	4	1	1	2
10	2	0	0	
Total	35	35		

(k) Differences (Balanced - Original)

Stop	Offs	Ons	Thru Load	Dep Load
1	0	1	0	1
2	0	0	1	1
3	0	0	1	1
4	-1	0	2	2
5	1	0	1	1
6	0	-1	1	0
7	0	-1	0	-1
8	0	0	-1	-1
9	1	0	-2	-2
10	0	0	-2	-2
Total	1	-1		

Target ons is calculated to be 27, target offs to 28. Corrections are distributed proportionally and rounded as before. Note that in balancing the late sub-block in Table 10(i), through load at Stop 5 is initialized to −1, the correction target.

Both segments, after balancing, have no negative load violations, so the correction procedure ends.

8.3.4 Other Count Correction Issues

Independent of the balancing procedure, several questions related to databases and corrected counts arise:

- **Should the database store corrected counts, or should corrections be made on the fly?** In the APC analysis systems studied, corrections are made on the fly. However, those systems used simple algorithms, not accounting for inherited passengers or loops. With more complex correction algorithms, it seems preferable to make corrections during entry processing, storing the corrected counts in the database.
- **Should raw counts be stored as well as corrected counts?** If corrected counts are stored, agencies and vendors both still want raw counts retained as well. Storing raw counts helps preserve the integrity of the data, is useful for investigations, and can be used for testing new balancing methods.
- **Can records with invalid counts be retained?** There are likely to be many trips for which count data is judged invalid, but for which time and location data, useful for analyzing operations measures such as schedule adherence, is not. The reverse can also be. This concern is met by including flags in the database for "counts invalid" and "times invalid."
- **Should corrected counts be forced to be integers?** Most users prefer that corrected counts be integers. The appearance of fractional counts in a database tends to raise questions and emphasizes that the raw counts were deemed incorrect. On the other hand, it is quite possible to store fractional counts and simply round analysis results.
- **Should the database store load, or should it be derived from counts?** Passenger load can be derived on the fly from counts starting at the beginning of a trip, provided there are records for inherited passengers. However, storing corrected load can be a way of compensating for not storing corrected on/off counts or inherited passengers, as is the practice at Tri-Met. Tri-Met uses uncorrected counts in analyses of boardings, while for analyses involving load it uses balanced load estimates.

CHAPTER 9

APC Sampling Needs and National Transit Database Passenger-Miles Estimates

Two common questions regarding APC system design are how much accuracy is needed from APCs and how many APC units are needed to obtain an adequate sample size. Answering those questions requires facing a related question: when an agency generates measures such as peak load, passenger-miles, and route boardings from APC data, how *timely* and *precise* must those estimates be?

9.1 Sample Size and Fleet Penetration Needed for Load Monitoring

According to the *Transit Data Collection Design Manual* (16), the passenger-count–based statistic requiring the greatest precision is average peak load on heavy-demand routes, a measure used to adjust headway. A reasonable target precision to ensure that the route is neither overcrowded nor overserved is 5% or 6%, effectively limiting permissible load bias on crowded segments to about 5%.

Sample size needed to achieve this target precision depends on the bias and *cv* of load estimates. Fleet penetration needed, in turn, depends on the number of daily trips on the route-direction-period being analyzed, the data recovery rate, and how the instrumented fleet is distributed. Fleet penetration of 10% will afford about $20r$ observations per quarter for a route-direction-period with five trips per day, where r is the data recovery rate. (For example, if $r = 80\%$, then $20r = 16$ observations that would be obtained.) If needed, greater sample sizes can be achieved by simply concentrating equipped vehicles on heavy-demand routes, at the expense of low-demand routes, for which less precision in load estimates is needed.

We have posited elsewhere that APCs make possible a more precise method of scheduling and service quality monitoring focused on extreme values of load rather than mean values. Extreme values reflect the impacts of load variability and service regularity as well as frequency and better reflect the quality of service as felt by passengers. Estimating extreme values requires a far greater sample size than estimating mean values, which is an argument favoring instrumenting the entire fleet with APCs, a course being pursued by Tri-Met.

9.2 Accuracy and Sample Size Needed for Passenger-Miles

All U.S. agencies receiving federal assistance and operating in urban areas are required to report annual systemwide passenger-miles by mode to the NTD. Traditionally, these estimates are made from a sample of manually counted ons and offs. Agencies can use a standard sampling and estimation procedure that requires on-off counts on 549 or more trips (39), or they can use any other sampling method that achieves a precision of ±10% or smaller at the 95% confidence level. Because manual on-off counts are labor intensive, there is a natural desire to find less burdensome measurement and estimation methods, including using APC-generated counts (40).

One factor in using counts measured by APCs is the accuracy of the counts themselves, which, as the previous chapter shows, depends not only on sensor accuracy but also on data processing techniques used for parsing, screening, and balancing. The second factor is having an adequate sample size. The two factors are related; the less accurate the counts, the larger a sample is needed. This section deals with that accuracy/sample size trade-off.

For all but the smallest transit agencies, sampling requirements for NTD passenger-miles reporting are considerably less demanding than are other uses of the data such as monitoring load or boardings by route, because the NTD precision requirement is only applied to a whole year's sample aggregated systemwide. Therefore, meeting the NTD requirement should be easy for almost any transit system with APCs. However, because the NTD requires that alternative sampling methods be statistically justified, the following section examines passenger-miles sampling and estimation with APCs in detail.

9.2.1 Standard Error Targets in the Presence of Bias

Let

$\bar{Y}$ = mean passenger-miles per trip
b = relative bias in the passenger-miles estimate ($b = bias/\bar{Y}$)
$\bar{y}$ = estimated mean passenger-miles
se = standard error of the passenger-miles estimate
$rse = se/\bar{Y}$ = relative standard error

The precision specification can be interpreted as:

$$P(\bar{y} - 0.1\bar{Y} \leq \bar{Y} + 0.1\bar{Y})$$
$$= P(\bar{Y} - 0.1\bar{Y} \leq \bar{y} \leq \bar{Y} + 0.1\bar{Y}) \geq 0.95 \quad (5)$$

Subtracting $E[\bar{y}] = \bar{Y}(1 + b)$ and then dividing by se,

$$P\left(\frac{-0.1\bar{Y} - b\bar{Y}}{se} \leq \frac{\bar{y} - \bar{Y} - b\bar{Y}}{se} \leq \frac{0.1\bar{Y} - b\bar{Y}}{se}\right) \geq 0.95$$

By the Central Limit Theorem, the middle term approaches a standard normal variate as sample size increases; therefore, using the notation $\Phi()$ = cumulative standard normal distribution, the precision requirement becomes

$$\Phi\left(\frac{-0.1 - b}{rse}\right) - \Phi\left(\frac{-0.1 - b}{rse}\right) \geq 0.95 \quad (6)$$

From relation 6, selected values of permitted relative standard error for a given value of relative bias are shown in Table 11. For manual data collection, assumed bias-free, the permitted relative standard error is 0.051; with 8% relative bias, the permitted relative standard error falls to 0.012. To be safe, a transit agency would do well to limit the permissible bias in passenger-miles or load to less than 8%.

Table 11. Relative standard error required versus measurement bias.

Measurement Bias*	Permitted Relative Standard Error*
0.00	0.0510
0.01	0.0500
0.02	0.0471
0.03	0.0423
0.04	0.0365
0.05	0.0304
0.06	0.0243
0.07	0.0182
0.08	0.0122
0.09	0.0061

*Relative to mean passenger-miles per trip

9.2.2 Sample Size and Coverage Requirements

The determination of sample size requirements assumes three stages of sampling: in stage 1 all routes are selected; in stage 2, for each route, certain timetable trips are selected; and in stage 3, for each selected timetable trip, certain days are observed. The assumed cv's for trip-level passenger-miles at stages 2 and 3 are:

$cv_2 = 0.9 = cv$ of timetable trip means (within route)
$cv_3 = 0.3 = cv$ of daily passenger-miles
(within a given timetable trip)

The assumed values are conservative estimates based on experience with data from many transit agencies. The values reflect the fact that, for a given route, most variation in trip-level passenger-miles is due to differences in where trips fall within the timetable (peak/off-peak, inbound/outbound), rather than random differences between days. Sample size requirements derived in this section are based on the weekday sample only; the addition of weekends, sampled with the same degree of fleet penetration as on weekdays, will improve precision, although not by much.

The effective penetration rate (f_3) is defined as the expected fraction of the daily schedule observed each day. It is the product of fleet penetration rate and data recovery rate.

Covering Every Weekday Trip

With an effective fleet penetration rate as small as 1% and careful rotation, every weekday timetable trip can be observed at least once per year. The annual estimate is determined by calculating average passenger-miles for each timetable trip, expanding by number of days that trip was operated, and summing over all timetable trips. Stratifying to this level is a very effective estimation technique because it eliminates the effect of variability between timetable trips. The weekday sample size requirement is

$$n \geq \max\left(N_2, (0.3/rse)^2\right) \quad (7)$$

where N_2 equals the number of weekday timetable trips and rse is the permitted relative standard error from Table 11. For bias up to 8% and for all but the smallest transit systems, the N_2 term will control; that is, it is sufficient to simply observe every timetable trip once.

Covering Most Weekday Trips (Two-Stage Sampling)

Logistics and data recovery problems can frustrate plans to observe every weekday timetable trip. The following plan

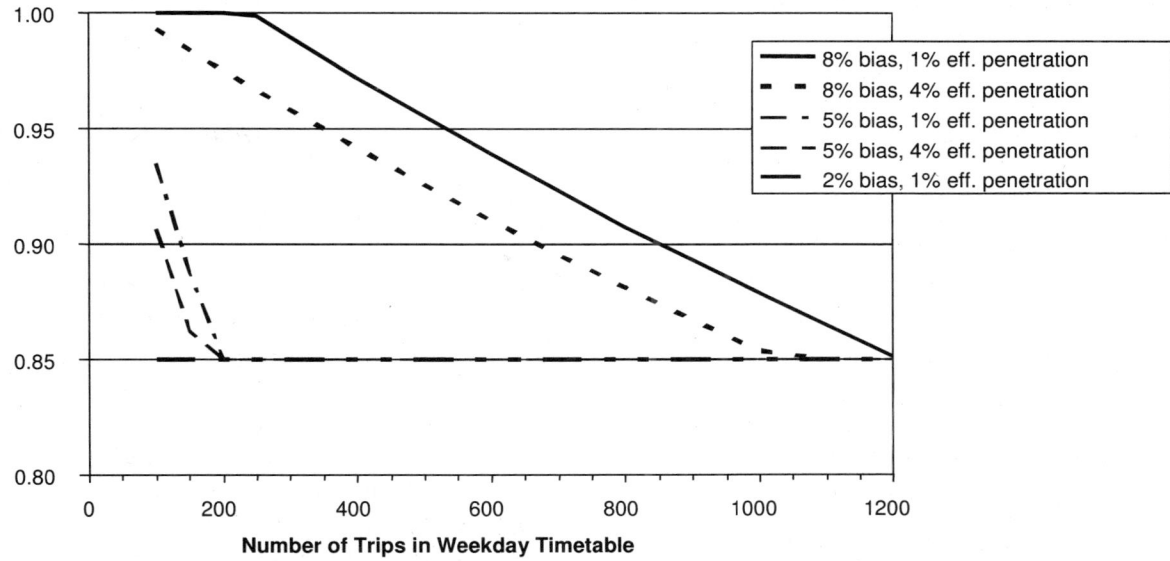

Figure 20. Timetable coverage rate required versus timetable size.

assumes that only a percentage (f_2) of the timetable trips is covered. The estimation procedure is to get an average for each timetable trip that was observed, determine the route average (per trip), expand each route average by the number of trips operated per year, and then sum over all routes.

The relative standard error of the estimate is given by

$$rse^2 = \frac{cv_2^2}{f_2 N_2}(1-f_2) + \frac{cv_3^2}{Df_3 N_2} \quad (8)$$

where D is the number of weekdays in the year (about 252). For all but the smallest transit systems, the third term will be insignificant, and the size of the relative standard error will depend mostly on f_2.

Using equation 8, Figure 20 shows the required timetable coverage f_2 versus the number of trips in the timetable (N_2) for selected values of bias and effective penetration rate. Degree of coverage is restricted to values of 85% or greater, because lower coverage rates suggest poor logistical management with likely sampling biases (e.g., whole routes being missed or seriously undersampled). With 5% bias, 85% coverage is sufficient even for an agency with only 200 trips in the weekday timetable and 1% effective penetration. Only smaller systems with moderate to large bias will need greater timetable coverage or effective penetration.

9.2.3 Intentional Sampling

The recommended estimation procedures just described involve unintentional sampling—the APCs collect data all year long, and the agency just rolls it up. This approach assumes that instrumented buses, for reasons beyond NTD passenger-miles estimation, are being circulated in a manner that covers the entire schedule regularly. Intentional sampling methods with limited sample sizes are clearly inferior, unless data processing procedures are still so undeveloped that each trip's data must be manually checked.

CHAPTER 10

Designing AVL Systems for Archived Data Analysis

One of the hard lessons learned is that off-line analysis has different data needs than real-time monitoring and that, therefore, AVL systems designed for real-time monitoring may not deliver the type and quality of data needed for off-line analysis. Considering the ways off-line data can be used to improve operations and management, and considering the way AVL system design affects what data is captured as well as its quality, this chapter presents findings related to AVL system design.

The emphasis of this chapter is on AVL rather than APC system design because APCs have always been designed for off-line analysis. To a large extent, this chapter summarizes Furth et al. (6).

10.1 Off-Vehicle versus On-Vehicle Data Recording

There are two options for recording data with AVL systems: on-vehicle (in an on-board computer) or off-vehicle (by sending data messages via radio to a central computer). As discussed in Section 2.3, the radio channel capacity limits an AVL system's ability to record data over the air. On-vehicle data storage is clearly superior in that it presents no effective limit on data recording.

Where radio-based systems are still contemplated, buyers must learn the capacity of any proposed system to collect timepoint and/or stop data, along with random event data. Capacity depends on the number of radio channels available, the number of buses instrumented, message length, and specifics of the technology used. The number of radio channels available to a transit system is strictly limited and varies by location, because radio channels are allocated by government. Radio-based systems have been successfully configured to record useful data for off-line analysis; for example, Metro Transit's system makes timepoint records from all buses and stop records from about 15% of the fleet by sending messages over the air to a central computer.

10.2 Level of Spatial Detail

As discussed in Sections 2.3.1 and 4.2.1, the choices in spatial detail of basic AVL records are polling records (occurring at arbitrary locations, when the bus is polled), timepoint records, and stop records. Collecting data at a finer level is also possible.

10.2.1 Time-at-Location and Location-at-Time Data

Polling data can be characterized as location-at-time data, giving bus location at an arbitrary time; stop and timepoint records, on the other hand, have time-at-location data, giving the time at which a bus arrives or departs from a specific location. Most off-line analyses, including analyses of running time and schedule adherence, require knowledge of departure time from standard locations. Therefore, stop and timepoint records are inherently better suited to off-line analysis of AVL data.

Theoretically, time-at-location could be estimated from polling data by interpolation. This method introduces interpolation errors, whose magnitude can be almost as large as the polling interval on segments in which the plausible bus speed has a wide range (e.g., because of intermittent traffic congestion). During periods of traffic congestion, it can be difficult with polling data to determine whether a bus reporting coordinates close to a stop is in a queue waiting to reach the stop, is at the stop, or has already left the stop and is waiting in a traffic queue.

The project survey did not find a single case of a transit agency routinely using polling data for off-line analysis except for playback to investigate incidents. Researchers used such a data stream from Ann Arbor for some operational analyses (41), but the process of going from raw poll messages to trajectories matched to route and schedule was too involved to become routine. The three case study agencies with round-robin polling data do not use it off line except for incident investigation using playback. Also, the survey indicates that all of the traditional AVL suppliers, even if they still use polling to support real-time applications, include timepoint records in their data streams as well.

However, a relatively new entry to the market, stand-alone "next arrival" systems, uses only polling data to track bus location. Like earlier AVL systems, they are designed around a real-time application and, according to the interviewed vendor, use polling data to minimize the amount of equipment installed in the vehicles, making such systems less expensive. This vendor claims to have obtained good test results using the data from its system for off-line analysis of on-time performance. The next arrival system's data stream includes predicted arrival time at stops (based on proprietary algorithms); as buses get close to a stop, predicted arrival time should become a rather accurate measure of actual arrival time, especially if the polling cycle is short, and therefore might be used as an approximation. A drawback of next arrival systems is that, while their application focuses on arrival time, most running time and schedule adherence analyses are concerned with departure time.

10.2.2 Timepoint versus Stop-Level Data Recording

Given that obtaining time-at-location data is important, what location detail is needed: stop level or timepoint level? Of course, stop-level data is needed for passenger counts; but, for operations analysis, what is the incremental value from getting data at all stops as well as at timepoints?

Because scheduling practice in the United States is based on timepoints, timepoint data is all that is needed for traditional running time and schedule adherence analyses. Metro Transit's AVL-APC system design emphasizes this distinction: on buses with APCs, stop records are created; while, on buses with only AVL, only timepoint records are created. Timepoint data tends to be favored by systems that rely primarily on radio transmission for data recording, because timepoint messages do not consume much radio channel capacity–timepoint messages are not very frequent and tend to be rather short, including only timepoint ID, time and location stamp, and identifiers. (Interestingly, this issue does not arise in the Netherlands, because almost every stop is a timepoint there. Also, stop spacing in the Netherlands tends to be about 60% greater than in the United States, resulting in fewer stops.)

However, stop-level detail offers advantages to a transit agency willing to go beyond traditional scheduling and operations analyses. Those advantages stem from (1) finer geographic detail for operations analysis and planning, and (2) the fact that stops are where customers meet the system, making stops a natural unit for customer-oriented scheduling and service quality analyses. In Section 4.4.5, several advantages of stop-level scheduling were cited, including better customer information (both off line, as part of trip planning, and in real time, for predicting arrival time), finer control, and ability to apply conditional (schedule-based) signal priority. Stop-level analysis helps enable operations analysis to identify points of delay, determine the impact of changes to stop location or traffic control, and analyze bunching. Stop-level data also permits more customer-oriented service quality analysis, such as enabling determination of passenger waiting time at stops.

With stop-level data, bus arrival and departure times are easier to determine at the end of the line. If the last segment's data is unreliable, what is lost becomes much smaller with stop-level data.

The practice of making timepoint, but not stop, records appears to be partly a relic of past practice, partly a limitation of radio-based data communication, and partly a simplification (for example, an agency with timepoint records only has to make sure its timepoints are mapped correctly, not all its stops). In today's technology age, with on-board data storage possible at relatively little cost, there seems little reason to settle for less than stop-level data.

10.2.3 Interstop Data

Automatically collected data on what happens between stops is not nearly as important as data about stops. However, there is nearly no marginal cost to making interstop records, which can support some useful applications. Examples include monitoring maximum speed (both as a check for speeding and as a measure of quality of traffic flow), monitoring time spent below crawl speed (as a measure of delay), and treating the bus as a GPS probe for mapping bus paths. Another possibly valuable use, mentioned earlier, is to analyze operations at terminals to help better determine actual arrival and departure times.

One possible configuration, applied at NJ Transit, permits records at regular, user-set intervals; for mapping a bus's path, the interval can be made quite small. Eindhoven's configuration, in which a record is made whenever speed crosses a crawl-speed threshold, can be generalized. By using a few different thresholds, users could estimate not only delay (time spent below crawl speed), but also a speed profile, which might be used to characterize traffic quality or to monitor speed in different speed zones. Tri-Met's configuration, in which only maximum speed between stops is recorded, is partly a concession to limited on-board data storage (which was an important factor in the mid-1990s). Given the current availability of low-cost on-board data storage, frequent interstop records can easily be accommodated. However, until now, the value of much interstop detail was not yet proven.

Frequent interstop records detail can aid in matching. For instance, speed records may help resolve situations such as when a bus stops twice at the same stop, is jockeying around at a layover, or holds (to avoid running early) away from a stop. NJ Transit is interested in using frequent interstop records for improving maintenance management by correlating operations measures with maintenance needs, particularly if future generations of its data collection system can integrate data from the vehicle drivetrain system.

10.2.4 Arrival/Departure Time Accuracy

AVL systems vary widely in the data captured with respect to arrival and departure time at stops and timepoints. Some off-line analyses need arrival time, some departure time, and some both. Therefore, an AVL system will be more valuable if it detects and records both arrivals and departures. Door sensors and the recording of door open and close events help improve the accuracy of arrival and departure detection as well. For example, suppose a bus stops two or more times in the neighborhood of a stop. Was the first the stop and the second simply traffic delay, or was it the reverse? Or did the bus open it doors both times, so that arrival time should be taken from the first stop and departure time from the second?

Without a door sensor, arrival is frequently detected by a vehicle entering a 10-m radius zone around a stop. Around major stops and terminals, the zone can be quite a bit larger, which can distort arrival time if a bus faces congestion getting to the stop (e.g., because a traffic queue or another bus is blocking the stop). Likewise, zonal detection can distort departure time if departing buses encounter congestion before leaving the stop zone, which can happen at near-side stops, at stops where buses have to await a gap or yield to crossing pedestrians before entering the traffic stream, and at terminals.

Knowing both when doors close and when the bus actually departs is valuable for detecting holding, which is important for running time analysis. Because of the possibility of holding, door close events are not sufficient to determine departure time. Therefore, while door sensors are valuable, they are not sufficient. (In many Latin American cities, where buses routinely operate with doors open, they are almost useless.)

10.2.5 Route Endpoint Identification

As mentioned earlier, many AVL systems are weak in determining when a bus arrives and departs a route terminal. For running time and schedule adherence analyses, these data items are critical, and system features that make their correct identification easier are valuable. Such system features include door open and close records, frequent interstop records in terminal regions, odometer-based records to supplement GPS-based records in terminal regions, and better algorithms for interpreting bus movements in terminal areas in order to better distinguish genuine departures from movements within a terminal area. When route terminals are located in zones with poor GPS reception (downtown or a covered terminal), data from supplemental devices and logic to interpret it are especially valuable.

10.3 Devices to Include

Integrating other devices in an AVL or APC system can add value either because the data those devices provide is inherently valuable or because of synergies that make the new data helpful for interpreting other AVL data.

Door sensors have already been mentioned for their value to help match location, determine arrival and departure times, and identify control time. APC systems virtually always include door sensors; their inclusion in AVL systems would be a benefit as well.

Odometer (transmission) data is helpful for determining bus speed, which can be valuable in its own right, and can be used to determine when a bus departs from a stop and when/where buses are delayed in traffic. Most AVL systems have odometer connections as a backup to GPS or to determine distance traveled between signposts.

Gyroscopes add richness to the data provided by odometers, allowing vehicles to be tracked off route and permitting matching based on turning locations.

Recording fare transactions in the AVL data stream is a means of getting location-stamped boarding data, which can be especially valuable to a transit system lacking APCs. When the payment medium is electronic and therefore offers an ID unique to the passenger, location-stamped fare transaction data offers the further opportunity for inferring link-trip and transfer information. However, there has been little experience to date with location-stamped fare records; this area has considerable opportunity for research and development.

Integrating the radio system's control head offers the opportunity to capture records of sign-in data, valuable for matching, and of operator-observed or -initiated events including pass-ups, special passengers (e.g., wheelchair and bicycle users, fare evaders), and traffic events (e.g., drawbridge up). These event records provide data that is valuable in its own right for direct analysis. Such records can also add detail and accuracy matching to running time and service analyses, for example, by helping to confirm and perhaps explain long delays or indicating when the "everybody boards the first bus" assumption behind waiting time calculations is violated. While radio-based systems always benefit from this connection, it would also benefit passenger counting and event recording systems. Also of potential value, but not yet applied (to the researchers' knowledge), would be coded records of standard radio messages initiated by the control center, such as instructions to hold for a connecting passenger.

Integration with the wheelchair lift (or lift sensors) would provide a more accurate and automated record of lift use than relying on operators to initiate a message.

Integration with the destination sign might prove useful to help with matching.

Integration with a stop announcement system or next arrival system does not bring in any new data; however, it creates an incentive for stop matching to be accurate and thereby benefits AVL-APC data analysis.

AVL systems have long attempted to use data from the vehicle's mechanical system (in addition to that from the odometer), such as oil temperature and air pressure. In real time, alarms from these systems have delivered so many false positives that they tend to be ignored. Whether the recording and off-line analysis of mechanical data integrated with location data can deliver new insights on mechanical performance is a rich area for further exploration.

10.4 Fleet Penetration and Sampling

AVL systems, when installed, are usually installed on the entire fleet. APCs have traditionally been installed on about 10% to 15% of the fleet.

Chapter 9 discusses how fleet penetration affects sample size, and what sample size need is for passenger count–related data items such as load, boardings, and passenger-miles. The general principle is that if the data is used only to determine mean values, small samples are sufficient; however, when extreme values are important, a complete or at least large sample is preferred. For passenger count data, 10% penetration is more than enough for boardings and passenger-miles data, for which only mean values are needed. For load on crowded bus routes, a near-complete sample is desirable so that extreme values can be observed. With a small fraction of the fleet instrumented, large sample sizes can still be obtained from crowded routes if the instrumented buses are disproportionately allocated to crowded routes.

On the side of operations data, only near-100% penetration will provide the large sample sizes needed to determine extreme values used in statistically based running time analysis and design. With a small fraction of the fleet equipped (as in a traditional APC system), large sample sizes can be obtained by aggregating over time, but at the risk of some of the data being out of date.

Headway analysis requires 100% instrumentation on a route at a given time, which can only be achieved with either 100% penetration or careful allocation of the instrumented sub-fleet.

Another benefit of full coverage is the ability to investigate complaints. As many complaints arise from extreme events (long waits, overcrowding), full coverage would be most helpful in such investigations.

Many transit agencies report that managing the allocation of an instrumented sub-fleet can be a large headache and that concerns other than data collection (e.g., who gets the new buses) often control the allocation, frustrating data collection plans.

A major motivation for instrumenting only a fraction of the fleet with APCs has been their cost. Tri-Met has shown that by integrating APCs with an AVL system, the incremental cost of an APC can be reduced to the $1,000 to $3,000 range; Tri-Met now treats them as standard equipment included in all new bus purchases. Note that Tri-Met uses a rather simple-technology APC and that more complex and (presumably) accurate APCs may not offer such an attractive incremental cost.

10.5 Exception Reporting versus Exception Recording

Exception reporting is certainly a valuable management tool, available for use with any AVL-APC data archive. It should be distinguished, however, from "exception recording," the practice of only recording a bus's location if it is off schedule or off route. This protocol of exception recording does not permit analysis of normal operations and should therefore be avoided.

CHAPTER 11
Data Structures That Facilitate Analysis

As mentioned in earlier chapters, original APC data consists chiefly of stop records, plus possible sign-in records. Original AVL data consists of stop or timepoint records, sign-in records, and records of various other events. It may also include polling records.

For analysis, these data records have to be screened and possibly corrected. Data that is not matched to a route and schedule should be matched. Beyond cleaning and matching, certain data structures may need to be created in the analysis database in order to facilitate analysis. Header and summary records offer some convenience for queries and analyses involving aggregation. Special data structures are needed to deal with multiple pattern analyses that are more than simple aggregations. Modularity in analysis procedures can also be enhanced by using standard, specialized database formats.

11.1 Analysis Software Sources

Software used in practice to analyze archived AVL-APC data can come from five different sources: in house, the AVL-APC vendor, a scheduling software vendor, a third party with a standard product, and a custom software developer. Each arrangement has its advantages and drawbacks.

11.1.1 Software Developed in House

Much of the current analysis of archived AVL-APC data uses home-grown software tools. This arrangement has worked well for some agencies, allowing them the flexibility to adapt to their particular needs and enterprise databases and ensuring that tool development is closely tied to need and likely use. For pioneering agencies, developing their own software was a necessity.

Since the mid-1990s, self-developed database and reporting software for AVL-APC data has used commercial off-the-shelf (COTS) database platforms on PC networks. COTS platforms have the advantage of being less expensive and benefit from regular upgrades, necessary in this age of technological advance. Coding for standard and ad hoc reports is prepared either in a database query language or using report-generating software such as Crystal Reports and Brio. Analyses that demand more complex calculations are often performed with spreadsheets or statistical analysis packages, with database queries used as a front end to select the data for analysis. One disadvantage of COTS database platforms and reporting software is that they can be slow when a lot of data is involved. Some agencies have found that powerful report-generating tools (available at 3 to 10 times the cost of their low-end counterparts) help overcome this problem by periodically pre-staging the data most likely to be used in reports and analyses. Response speed for large datasets can also be reduced by use of special data structures optimized for fast data retrieval.

Tri-Met is an example of an agency whose AVL-APC data analysis software was developed in house. Data is stored and managed in an Oracle database. Using a query language, selected data (e.g., by route, direction, times, dates) can be extracted. Extracted data is then imported to a commercial statistical analysis system (SAS) for numerical analysis. Scripts for standard queries and analyses are stored and reused. Sometimes results are imported to Microsoft Access for easier formatting.

King County Metro, with separate AVL and APC systems, uses multiple databases and applications. Its AVL data is stored in an Informix database. For schedule deviation analysis, scripts coded in Microsoft Access provide a friendly user interface for selecting AVL route, direction, time, date range, and so forth. The analyses themselves were programmed in a query language and are performed by Informix, which produces output in the form of Microsoft Excel tables and graphs. Analysts may do further manipulations of the Excel tables. For running time analysis, a query language program runs every 2 weeks on the AVL Informix database, extracting data that is then input to their scheduling package, Hastus, which includes the add-on software product ATP for running time analysis. Raw APC data is kept in an Oracle database. Using programs prepared in the Focus query language, summary records are created and exported to a Microsoft Access database, which has been programmed to offer a friendly user interface and nice reports. There are also standard reports created using query language from the original databases.

Metro Transit, a third example, analyzed running time data from its now obsolete AVL system using macros written in Microsoft Excel, once the analyst had extracted the data of interest from the database. In its new AVL system (now in implementation), Metro Transit is working with the AVL vendor to define analysis and reporting needs; they plan to share responsibility for development of analysis software.

Two final examples are NJ Transit and Broward County Transit, whose APC/event recorder and AVL systems (respectively) are operational and expanding. They are using COTS database platforms for data management and COTS report-generating software Brio and Crystal Reports for analysis.

Unfortunately, developing one's own database and reporting software demands resources and expertise that are beyond the reach of many transit agencies. Because of differences in software platforms and data formats, tools developed at one agency are usually not transferable to another.

11.1.2 Software Supplied by Equipment Vendors

Software supplied by some APC vendors provides useful reports including on/off/load profiles, running time distributions, and on-time performance. However, it usually lacks flexible query capabilities.

Historically, AVL vendors provided software related to real-time applications only; for archived data analysis, their job ended when they handed the transit agency the data. Often, the only archived analysis tool is the ability to play back the AVL data stream. Some AVL suppliers include a genuine database and analysis function, but tend to offer only elementary analyses such as on-time performance percentages and reports on how often various event codes were transmitted. For two of our case study agencies, AVL vendors are developing more comprehensive database and analysis capabilities as part of their procurement contracts.

Software that is coupled to on-board equipment limits the flexibility to add other on-board equipment or to replace aging equipment with equipment from a different vendor. The vendor may go out of business or may not continue to improve the software. Furthermore, a note of caution comes from reviewing 20 years of industry experience with farebox data. While the major electronic farebox vendors also supply software for analyzing farebox data, most larger U.S. agencies who rely on farebox data for monitoring ridership have found that they had to export the data to a database developed in house to run their own reports because the vendors' software did not provide the flexibility they needed.

11.1.3 Software Supplied by Scheduling System Vendors

Analysis programs offered by scheduling system vendors focus on analyzing running time data to suggest scheduled running times. An example is the tool used at King County Metro. Because it is tied to the scheduling system, its suggested scheduled running times can be semi-automatically entered into the scheduling system database. Ironically, for the version seen in the 2002 case study, its running time analysis is performed without reference to scheduled departure times or headways and, therefore, cannot analyze schedule or headway adherence, or report results for particular scheduled trips.

Software coupled to the scheduling system has many of the same disadvantages as software coupled to an equipment vendor. One case study agency that uses such a tool for running time analysis has to use its own database and software tools for other analyses and ad hoc queries.

However, one advantage of this source of software is that for scheduling system vendors, software development is their business. If they take on AVL-APC data analysis seriously, they are well positioned to develop some very good tools and to maintain them. With many customers worldwide, they are in a good market position if they choose to exercise it.

11.1.4 Third-Party Software

In the Netherlands, Delft University of Technology's Transportation Engineering Laboratory has developed the database and reporting software TriTAPT for detailed analyses of AVL and APC data. Various editions have been applied over the last 20 years to several Dutch transit agencies; the current edition is being used in Eindhoven and in the Hague. It features many useful single-route reports; excellent graphical representations, including proportional scaling to represent distance and time intervals; attention to distributions and extreme values as well as mean values; a graphical user interface for selecting days and times to be included in an analysis; edit capability that allows an analyst to suppress outliers; and practical tools for suggesting scheduled running times. It has been applied with data gathered using a variety of automated data collection equipment, including APCs, event recorders, and AVL systems of different makes. Interfaces have been developed to scheduling system databases. It uses a custom database to speed processing, but includes an export and import utility so that data tables can be transferred to and from text files.

In Germany, the Hannover transit system Uestra developed its own database and reporting software for AVL data; a related spin-off company has recently commercialized it.

In the United States, to the researchers' knowledge, the first application of third-party software for analyzing AVL data is under way at CTA for use with a new smart bus system that features stop announcements and event recording on all buses and passenger counting on a sample of the fleet. Called RideCheck Plus because its analysis reports were originally developed for ride check data, it includes many standard analyses (e.g., load profiles and schedule adherence) and also offers some GIS capabilities including links with demographic data and mapping.

Third-party software for analyzing archived AVL-APC data has the advantage of modularity, not being tied to any particular brand of equipment or scheduling system. As a stand-alone product, it is likely to continually improve, unless the product is discontinued. It offers the benefits of standardization and replication. A major disadvantage of third-party software in the United States is that transit agencies' funding mechanisms often forbid them to buy software only for data analysis, although such a purpose often can be justified within the context of a major AVL or APC system procurement as was the case at CTA.

11.1.5 Software Developed for Custom Analysis

Several specialized AVL analyses by university research teams have been reported in the literature, including the previously mentioned analyses done at the University of Michigan using Ann Arbor Transit Authority AVL data and at Morgan State University using (Baltimore) MTA data. In both of these cases, the specialized processing required to analyze these datasets left them inaccessible to staff analysts. In contrast, Tri-Met's APC database, developed in house, supports analyses by both staff analysts and researchers from Portland State University.

11.2 Data Screening and Matching

As AVL and APC data is retrieved, it usually undergoes some "entry processing" before being entered into the archive database. Entry processing involves screening for and perhaps correcting errors. If data is not already matched, entry processing includes matching data to the schedule and base map. Data that cannot be matched, or is rejected in the screening process, is logically rejected (usually, not by discarding the data, but by flagging it as unusable). Some AVL-APC databases have flags indicating "don't use counts" (for passenger counts that were rejected) and "don't use times" (for invalid time data).

Screening involves typical checks for consistency and range. For example, passenger count data will be rejected if on and off totals for a vehicle block differ too much.

While processing AVL-APC raw data is usually automated, daily monitoring by a skilled analyst is valuable, at least during the break-in phase of a system, to see what data was rejected and why. Failure patterns can indicate a need for on-board equipment to be repaired or adjusted or for the base map or schedule to be updated. Some transit agencies have developed semi-automatic correction processes. For example, Houston Metro's screening program checks operator and run codes against the dispatch database; if a small discrepancy is found that could be explained as a simple keying error, it is either corrected automatically or brought to the attention of a person monitoring the process who can make or authorize the correction.

11.2.1 Full and Partial Matching

A large part of entry processing is checking location and time stamps for a match against the schedule and base map. A fully matched record will indicate the stop or timepoint ID, the scheduled trip ID, and the scheduled departure time from the stop or timepoint. Matching to stop ensures that records will be analyzed in the right sequence, and matching to scheduled departure time allows analysis of schedule adherence and selection of trips based on scheduled departure time.

In some AVL-APC systems, stop and timepoint records are already matched to stop and scheduled trip; processing simply checks for consistency. Other AVL-APC data streams have to be matched during entry processing. For example, Tri-Met's AVL data records have only vehicle block ID, with time and GPS coordinates. Matching correlates GPS coordinates with stops, parses trips, and adds trip ID and scheduled departure time fields from a table correlating trips with blocks. When the vehicle block is known, tracking is much easier.

For stop records, matching can include checks for whether consecutive stop records should be merged, as when a bus closes its doors and advances a few feet, but then reopens its doors to let some more passengers in or out. In Tri-Met's entry processing, multiple stop records for the same stop are not directly merged; rather, a flag indicates which records are "primary" (the first stop record for a given stop) versus "secondary." Calculation routines are programmed to logically merge secondary stop records with their primary record. Stop or timepoint record processing may also involve inferring arrival or departure time by adding or subtracting a constant travel time from the recorded time, when the recorded time occurs a known distance from the stop or timepoint.

In many AVL systems, a bus passing a stop or timepoint without stopping will cause a stop or timepoint record to be generated on board. If not, records for stops that were skipped can be generated as part of the matching process, as is done at Tri-Met.

If polling data were to be used for more than playback analysis, matching would be done as part of entry processing to create stop records from it.

In some analysis databases, records are only partially matched. For example, they may indicate the timepoint, but not the scheduled trip. This kind of record can support many types of analysis, such as analysis of running time between timepoints. However, without matching to scheduled trip, control time cannot be inferred because it becomes impossible to know whether a bus was running early. (In principle, one could analyze schedule adherence by comparing the array of scheduled departures with the array of observed departures; however, the reality of missing data makes a simple comparison impractical.)

Including a field for scheduled departure time enables selection based on scheduled times. Without scheduled departure time as a field in a stop or timepoint record, one can select data for analysis based on a range of actual departure times (e.g., analyze all the trips that began between 7:00 a.m. and 9:00 a.m.); that kind of analysis is often done for running times. A disadvantage of selection based on actual running times is that the set of trips included on any given day can vary depending on whether trips near the period boundary departed before or after the period boundary; such variation in the numbers of trips included in a day's analysis can distort results.

11.2.2 Trip Parsing

Matching also involves parsing the data stream for a given block/day by trip. Many of the issues involved in identifying trip endpoints have been discussed in Section 2.2.5. Parsing passenger counts at trip ends is discussed at length in Chapter 8. One common parsing operation is converting a single record indicating the end of one trip and start of another into two records, one for each trip.

11.2.3 Trip Header Records

Entry processing can involve the creation of header records for trips and blocks. (Trip summary records, which serve a different purpose, are described later.) Header records, which are part of TriTAPT's data structure, help organize the database and make selection quicker. The header record for a trip contains pointers to that trip's stop records, as well as trip-level information such as route ID, trip ID, and scheduled and actual trip departure times. These header records make queries faster, as the query only has to determine which trip headers meet the selection criteria. Many databases, including Tri-Met's, function without trip headers, including all the header information in each stop record. Queries directly select stop records, which can make queries slower in a large database.

For a trip header, an alternative (or supplement) to scheduled and actual departure times from the start of the trip is departure or arrival time at a designated key point, which may be different from the starting point. On radial routes, the time at which a trip enters or leaves the downtown may be a more meaningful choice for categorizing it by period than the time it began; this distinction can be especially important if a system has a mix of short and long routes.

As mentioned in Chapter 8, several transit agencies are hoping to move to trip-level passenger count screening and correcting as part of entry processing. If that is done, the number of inherited and bequeathed passengers determined for each trip can be incorporated as fields in the trip header records.

11.3 Associating Event Data with Stop/Timepoint Data

For most routine analyses of AVL-APC data, the fundamental record type is the stop or timepoint record. However, several analyses involve data from other types of event records or from interstop records. Examples include information about pass-ups or wheelchair lift use (which occur at stops but may be recorded as a separate event), and maximum speed or drawbridge delay (which occur between stops). One database issue is how to associate data on those kinds of events with stop or timepoint records so that they can play a part in passenger count or running time reports, for example.

11.3.1 Adding Fields to Enrich Stop or Timepoint Records

One way of associating the information contained in other types of records with stop records is to add to stop or timepoint records fields summarizing information from other record types. As an example, stop records in Eindhoven's TriTAPT database include fields for segment delay and control (holding) time. Segment delay is calculated as part of the entry processing of AVL data, using records of buses crossing a crawl speed threshold to calculate the amount of time between a pair of stops spent stopped or below crawl speed, excluding time spent at the stop. Likewise, control time is calculated based on whether a trip was early and how long its dwell time was. Because each TriTAPT stop record contains information on both a stop and the segment following it, it is called a "stop module" record.

In Tri-Met's database, a field for maximum speed achieved on each preceding segment is part of the stop record. In fact, this field is filled on board when the stop record is created, rather than as part of entry processing.

Because AVL systems in the United States record a variety of event types that can be relevant to operations analysis, part of this project involved making the structure of the TriTAPT database more flexible, allowing an agency to include any number of fields in a stop record. Examples are numeric fields for maximum speed and binary fields for whether a particular event type (e.g., pass-up or drawbridge delay) occurred at the stop or on the segment following.

Incorporating interstop summaries in the stop record provides adequate geographic detail for many purposes. Where an interstop segment does not provide adequate geographic detail (e.g., if there are two traffic bottlenecks between stops and the delay at each bottleneck needs to be identified), analysts can simply add a dummy stop to the base map.

If the database's fundamental record is a timepoint rather than stop record, the length of a timepoint segment creates a considerable loss in geographic detail if events that occur at stops and en route are simply labeled as occurring on a timepoint segment. For some analyses, however, this loss of detail is unimportant. For example, in a running time analysis, it may be sufficient to know how often the bicycle rack or wheelchair lift is used on each timepoint segment; where on the segment it was used does not matter. However, if it does matter, one could query the original event records.

11.3.2 Matching Other Record Types

An alternative to incorporating summaries into stop records is to associate each event record with a stop (either where the event occurred or the last stop visited for en route events) and departure time, just like stop records are matched. Tri-Met follows this approach, adding to event records fields indicating the nearest stop and distance from that stop.

Analyses that want to merge stop record information with information from other event record types can select multiple record types and use the stop and scheduled trip as keys to correlate records. Of course, that kind of on-the-fly merging of data from multiple record types is more complex and time consuming than one in which the data was merged during entry processing, but it is also more flexible. If event records are not labeled with a stop or timepoint, matching and merging them on the fly with stop or timepoint records would be impractical.

From the survey, the use of event records other than stops and timepoints appears to be only on an ad hoc, analyst-intensive basis. For example, seeing an unusually large running time might prompt an analyst to query whether there was an event that caused a major delay on a segment or a specialized study might query bicycle rack events to get an idea of where they occur. However, to the researchers' knowledge, bicycle rack events and similar event data are not part of routine running time or demand analysis tools.

11.4 Aggregation Independent of Sequence

Almost all analyses other than incident investigation involve aggregation: over multiple days of observation, over multiple stops or segments, over multiple scheduled trips in a day, over multiple patterns that make up a line, or over multiple lines.

An important distinction in aggregation is whether an analysis has to follow a sequence of stops or trips. In many analyses, stop and trip sequence are irrelevant; once the appropriate stops and trips have been selected, the result is a simple aggregation. Examples include total ons; maximum load; and number of timepoint departures that are early, on time, and late. Summary measures that do not involve calculations along a sequence of stops can easily be summarized over multiple patterns and multiple lines and lend themselves also to comparison between lines.

11.4.1 Summary Records for Routine and Higher Level Analyses

Transit agencies often have certain routine analyses that involve this simple type of aggregation. To reduce processing time, summary records can be created at the trip level, containing such items as total ons; maximum load; and number of timepoint departures that are early, on time, and late. An analysis such as average or distribution of boardings per trip on a route, or percentage of early/on-time/late departures, can be performed using those trip summaries. Higher level summaries (e.g., aggregating over a week or month, or over a period of the day, or both) can speed processing for reports needing only summaries at that level, such as quarterly route performance reports and historical trend analysis.

At higher levels in a transit agency, reports using AVL-APC data often involve data from other sources as well, such as data on revenue, accidents, or customer satisfaction. This kind of report is best generated by a general management database. The AVL software's responsibility is to create summary records that can be exported to the general management database, which also can be used for comparison reports, historical trend reports, and other such higher level reports. At Brussels' transit agency, for example, the AVL system generates line-level summaries of schedule adherence and passenger waiting time for every 2-week period; those summaries are exported to the general management database that is used to analyze route performance along many dimensions. Of course, this arrangement requires a well-developed enterprise database to receive the AVL summary records.

Planning analyses, including those that use a GIS, generally want to use long-term average passenger count, running time, and service quality data. AVL-APC systems can supply those averages and export them to the planning/GIS database.

When summary records are created, it is still important to preserve the possibility of drilling down to original records that have not been aggregated.

11.4.2 Accounting for Varying Sample Size

The number of days each scheduled trip is observed in a given date range can vary because of imperfect data recovery, especially if data collection uses a rotating instrumented subfleet. Analyses should account for these varying sample sizes by aggregating in a way that gives every scheduled trip, not every observation, equal weight.

For example, an easy but incorrect way to determine on-time performance for a line over a date range is to query all the timepoint records that qualify, and simply get a total of the number with early, on-time, and late departures. However, if some trips were observed more than others, such an estimate will be biased in favor of the trips with higher sampling rates. The proper estimation method would be, first, to get an average number of early, on-time, and late departures for each scheduled trip by aggregating over observed days and, then, to sum over all the scheduled trips that qualify.

If the sample size is so small that some scheduled trips were not observed, an alternative aggregation scheme is to aggregate over observed days within short periods (e.g., 1-hour periods), then expand each period's result according to the number of scheduled trips in that period, and aggregate over trips.

11.4.3 Accounting for Missing Data

Two approaches may be taken to deal with trips that were not observed on a given day. The classic approach is to omit them from the dataset and to give analysis algorithms appropriate methods to deal with missing data and account for the varying sampling rates that result, as discussed previously.

An alternative approach is to place imputed values into the database whenever data is missing. Imputed values may be based on historical averages or on values from "similar" trips. That approach allows analysis algorithms to not have to deal with missing data or varying sampling rates. However, supplying imputed values is a controversial practice that, to the researchers' knowledge, has not been done with AVL-APC databases.

11.5 Data Structures for Analysis of Shared-Route Trunks

Analyses in which stop sequence plays a role are generally called "profiles," showing results along the route. Examples are load profiles, running time profiles, delay profiles, and profiles of schedule deviation and headway irregularity. Creating a profile requires an unambiguous stop sequence, generally provided in a stop list for each pattern (sometimes called "branch" or "variation"). A trip that deviates by even a single stop must be classified as a separate pattern.

Profiles can readily be aggregated over scheduled trips following the same pattern. Aggregating this kind of report over completely different patterns is meaningless. However, a certain pattern that falls between "same" pattern and "completely different" pattern presents an analysis challenge. Many transit systems have route structures in which a line consists of multiple patterns (cases of up to 20 patterns have been observed) that share a common trunk. When several patterns share a common trunk, analysts might be interested in the load profile over the trunk, in an analysis of headways along the trunk, or in an analysis of running times or delays for all patterns along the trunk.

In the survey of practice, the only shared-trunk analysis capability seen was for running time, in which running time was analyzed for all trips making a selected sequence of timepoints. Methods to analyze headways and load profiles on a trunk were either non-existent or ad hoc (i.e., applicable only to the particular trunk for which they were developed).

As part of this project, a data structure for trunks was developed and tested in the TriTAPT environment. Users can define a "virtual route" consisting of a sequence of stops that may be shared, entirely or in part, by multiple route patterns. Users specify the patterns that contribute to the virtual route, specifying at which stop those patterns enter and leave the virtual route. A pattern may enter and leave a virtual route more than once. That way patterns that deviate from a main route (e.g., to serve a school or senior housing development for a few trips each day) can be accommodated. Load, schedule adherence, headway irregularity, and delay profiles along the virtual route will then reflect all the trips on the trunk, including patterns that branch off it or take detours.

The virtual route pattern is stored as a permanent data structure, and any analysis that can be performed on a single route can be performed as well on the virtual route; in the latter case, all trips belonging to route patterns that contribute to the virtual route are queried for the analysis. Load profiles made for virtual routes have to account for passengers already on board when a trip enters the trunk.

11.6 Modularity and Standard Database Formats

As mentioned earlier in this chapter, analysis software developed by a third party offers modularity and the possibility for analyzing AVL-APC data without developing one's own software. However, using third-party software requires using a standard data structure, which may in turn demand routines to convert data from its native format. That approach is working in practice for the transit agencies in Eindhoven and the Hague that use TriTAPT and for transit agencies in both North America and Europe that use running time analysis tools provided by scheduling software vendors.

As part of this project, the ability to interface archived AVL-APC data from North American transit systems to the TriTAPT data format was tested. Conversion routines were developed successfully for three U.S. transit agencies and one Canadian transit agency, all having different native data formats. The Delft University of Technology has made TriTAPT conversion routines publicly available, allowing agencies to select one that starts with a database similar to theirs and modify the program as necessary.

In principle, agencies should be able to use a third party's analysis routine yet customize the reporting format. Besides cosmetic changes (e.g., inserting a logo), agencies might wish to make substantial changes in how results are formatted.

This desire can be accommodated by having analysis routines export their results as simple tables, which agencies can then import and format as they wish, perhaps using report-writing software. For example, all of TriTAPT's analyses, in addition to generating a standard graphical report format, also generate a table containing all of the numeric results that can be readily exported to a database, spreadsheet, or other platform for formatting as the agency desires.

CHAPTER 12

Organizational Issues

Effectively capturing, archiving, and using AVL-APC data involves overcoming organizational as well as technological challenges. This chapter summarizes the chief challenges reported by the surveyed agencies.

12.1 Raising the Profile of Archived Data

A reason that many AVL systems have failed to deliver their potential in terms of useful archived data is that those who specified and designed the systems either did not emphasize the importance of archived data or, more likely, did not recognize important differences between the needs of real-time data and those of archived data (*38*). Time and again, procurements have focused on real-time applications, with the implicit expectation that archived data analysis would somehow happen. Some of the lack of appreciation of the character and value of archived data has been on the part of vendors whose primary product is real-time information; and some has been on the part of transit agency staff who managed procurements. There is a need for transit agency staff who are involved in system procurement to better understand how system design affects what data is captured, what the data quality will be, and what off-line analyses it will be able to support. At the same time, there is a need for decision makers to appreciate the importance of archived data acquisition and analysis for improving system management and performance.

Several studies have looked for quantifiable benefits of AVL systems to justify their cost. Where benefits have been quantified, they most often come from an off-line application, namely, revising scheduled running times. This is ironic, considering that off-line analysis has often been an afterthought of such systems.

12.2 Management Practices to Support Data Quality

Control and supervision has traditionally been concerned about performance, not about data collection. In fact, to the extent that buses are ordered to deviate from their schedules, operations control makes schedule matching more difficult.

Fortunately, agencies are becoming more aware that collecting and later analyzing data can also contribute to improving performance. There is an opportunity to improve data quality by changing control and supervision practices. Examples include detecting and correcting invalid sign-on data, informing the AVL-APC system of a revised schedule when a bus is deviated from its schedule, and standardizing codes for control messages.

12.3 Staffing and Skill Needs

To date, transit agencies making good use of AVL-APC data have been able to do so only because of the strong set of staff skills they have been able to employ. The lack of available data analysis tools has meant that agencies have needed the skills to develop their own database and analysis tools. Only because of dedicated, qualified, and resourceful staff were Tri-Met and King County Metro able to make the great strides they did in improving the quality of their AVL-APC data and in developing tools to analyze it.

In the future, with the development of third-party analysis software and increasingly relevant and accepted standards for data definitions and interfaces, the need for expertise in information technology (IT) to develop archived AVL-APC data systems should decline. However, agencies will still need the staff and expertise to analyze the data.

Managing data quality takes considerable staff effort. At agencies with good AVL and APC data, one or two staff members are usually devoted to overseeing that the systems deliver the data they should. Identifying and correcting accuracy problems sometimes takes considerable IT expertise, especially if matching algorithms or data objects have to be changed. With AVL systems, matching, accuracy, and data capture issues are often just as much a problem for real-time applications as for archived data, so that little additional work is needed for archived data itself.

12.4 Managing an Instrumented Sub-fleet

Where APCs are used, common practice is to equip only 10% to 15% of the fleet. The logistical issues of managing an equipped sub-fleet are significant. A program manager is needed to make assignments of instrumented buses and to check whether assignments were made. Garage supervisors must stage the instrumented buses properly, and transportation supervisors must ensure they are used where assigned. Data collection efforts that require all the buses operating on a route or along a trunk be instrumented at the same time (e.g., for a headway analysis) are particularly demanding; if a few instrumented buses miss their assignments, the data collection effort will have to be repeated another day. Another layer of complication is added if so many instrumented buses are needed in one place that intergarage transfers are needed.

Aging of the equipped sub-fleet also poses problems. At first, the instrumented buses are new, and there may be political demands to assign the new buses in ways that conflict with a data collection program. As they age, restrictions on using them on certain routes or runs can develop. For example, they may not have the ergonomic seats needed by some operators, or the low floor required on some routes. They may develop maintenance problems with wheelchair lifts or other bus systems.

Partly to avoid the complications of shifting instrumented buses around, Tri-Met is well on its way to equipping its entire fleet with APCs. With 65% of the fleet instrumented with APCs (and the entire fleet instrumented with AVL), Tri-Met simply allows the APC-instrumented buses to collect data wherever they get assigned; that gives it an adequate sample size. When APCs can be integrated into a smart bus system that already includes AVL and schedule matching, the marginal cost of adding APCs drops dramatically.

12.5 Avoiding Labor Opposition

Suspicion of "the spy in the cab" or "big brother" is natural. There is always the danger of sabotage if transit operators resent the way a system monitors them, especially if they believe that the system is unfair or inaccurate.

For the most part, transit agencies that have adopted AVL and APCs have avoided incapacitating labor opposition. Generally, agencies have been successful at communicating the security benefits of AVL, which builds operator support. APCs generally do not engender opposition, perhaps because their name suggests they are only counting passengers. Some agencies intentionally avoid directly challenging an operator with AVL-APC data. However, some use the data to identify patterns of abuse or poor performance, alerting supervisors where and when problems are likely to occur.

CHAPTER 13

Conclusions

Archived AVL-APC indeed holds great potential for improving transit management and performance. This report has reviewed the history and state of the practice in AVL-APC data collection and analysis. System design affects in many ways the type and quality of data captured, which in turn affects the types of analyses that the data can support. To develop guidance for system design, existing and potential analyses and tools that use AVL-APC data to improve management and performance are reviewed and their data needs analyzed. Analysis tools for running time, waiting time, and crowding were developed in the course of this project.

AVL systems have traditionally been designed primarily for real-time applications. For investigating specific incidents, archives of almost any AVL data stream can be used in playback mode. However, for analysis of historical data combining multiple days of observations, data requirements substantially exceed what will suffice for real-time monitoring. The following are the main conclusions regarding how AVL-APC systems should be designed to provide a valuable data archive:

- Storing data on board frees the system from the capacity restrictions of transmitting data records over the air.
- Time-at-location data (i.e., stop and timepoint records) is needed for analyses that aggregate over multiple days of observation, such as running time and schedule adherence analysis. Location-at-time (i.e., polling) data is not suitable.
- Stop-level records permit running time analysis and schedule making at greater geographic detail than timepoint records, better serving passenger information needs and supporting better operational control.
- Integrating on-board devices adds information to the data stream that can be valuable in its own right as well as aid in matching captured data to the base map and schedule. The most valuable devices to integrate are door sensors, odometer (transmission), and radio control head.
- Designers should pay attention to a data collection and processing system's ability to accurately determine arrival and departure times at stops, identify holding, and deal with multiple apparent stops and starts at bus stops and in terminal areas. Having data from both door sensors and odometers is particularly valuable in this respect.
- Integrating AVL with the fare collection system offers a potentially powerful means of measuring ridership patterns, because matching fare media serial numbers offers a means of observing linked trips.

Automatic data collection can revolutionize schedule planning and operations quality monitoring as agencies shift from methods constrained by data scarcity to methods that take advantage of data abundance. The large sample sizes afforded by automatic data collection allow analyses that focus on extreme values, which matter for schedule planning (e.g., how much running time and recovery time are needed, what headway is needed to prevent overloads) and service quality monitoring (e.g., how long must passengers budget for waiting, how often do they experience overcrowding). Stop-level data recording provides a basis for stop-level scheduling, a practice with potential for improved customer information and better operational control. With AVL-APC data, trends can be found that might otherwise be hidden, such as operator-specific tendencies and sources of delay en route. Regularly analyzing AVL data gives a transit agency a tool for taking greater control of its running times by offering a means of detecting causes of delay and evaluating the effectiveness of countermeasures.

Two sets of analysis tools were developed as part of this project. One uses running time data to suggest periods of homogeneous running times, analyze user-selected running time periods and scheduled running times, and create stop- or timepoint-level schedules. It includes a valuable "what-if" tool that allows schedule planners to propose a scheduled running time period and running time, and immediately see how that running time would have performed based on the historical data. The tool offered for segment-level running times uses a statistical approach that, if combined with operational control in the form of holding early trips, has great potential to improve on-time performance. These running time tools are part of the software package TriTAPT, developed at the Delft University of Technology, and are available with no license fee to U.S. and Canadian transit agencies through the end of 2009.

New tools were also developed on a spreadsheet platform to evaluate waiting time and crowding from the customer's perspective using AVL and APC data. Unlike traditional methods, they focus on the extreme events (e.g., very early and late buses, very long headways, very crowded buses) that most affect customer satisfaction. A whole new framework was developed for evaluating passenger waiting time, one that gives attention to the time that passengers have to budget for waiting, not just the time they actually spend waiting. Three new measures of waiting time are proposed: budgeted waiting time, potential waiting time, and equivalent waiting time, the latter being a comprehensive summary of passengers' waiting cost. This framework is superior to traditional measures of waiting time because it accounts for the impact of service unreliability on passenger waiting time.

The new crowding measures developed are defined from the passenger rather than vehicle perspective. With APC data, the percentage of passengers at the maximum load point who sit and who stand can be inferred, and they can be further divided into those who do and do not sit next to an unoccupied seat, and those standing at various levels of crowding. The result is a distribution of passengers by crowding experience, something that should correlate well with passenger complaints and with service objectives.

Routine use of automatic passenger counts poses a special challenge because imperfect counting accuracy requires trip-level parsing and balancing both for consistency and to avoid drift errors. Accuracy measures are described and analyzed. The researchers show, both from empirical data and from logical considerations, that accuracy in measured load (and, by extension, in passenger-miles) can be significantly worse than accuracy in on-off counts; therefore, load accuracy should be an important consideration in system specification, design, and testing.

Shortcomings in several existing methods of load balancing are pointed out; some of them can bias load and passenger-miles estimates upwards. A method of trip-level balancing is presented that prevents not only negative departing load, but also negative through load, a stronger feasibility criterion.

Approaches and data structures are described for dealing with end-of-line passenger attribution, especially on routes ending in loops and on interlined routes in which passengers can be inherited from one trip to the next. On routes ending with short loops (short enough that it can be assumed that no passenger trips both begin and end in the loop), a parsing and balancing method was developed that makes the loop effectively serve as a zero-load point. On routes lacking a natural zero-load point, such as downtown circulators or routes with loops on both ends, APC systems may need operator input to fix the load at a key point on each round trip.

The typical fleet penetration rate (10% to 15%) for APCs is shown to be adequate for all passenger count applications except for monitoring extreme crowding. The sample sizes typically afforded by APCs are shown to be sufficient to satisfy NTD requirements for passenger-miles reporting. Statistical requirements on systematic error in load or passenger-miles measurements to meet NTD reporting requirements are also given.

Analysis procedures using AVL-APC data should account for variable sampling rates by aggregating and weighting observations based on the schedule. Simply aggregating over all the stop or timepoint records for a chosen route, period, and date range can bias results in favor of trips that were measured more frequently.

While some analyses involve simple aggregation over a selected set of records, others require that analysis follow a particular sequence of stops. For such an analysis to involve data from multiple patterns operating on a common trunk, a special data structure is required to align the patterns. Such a data structure, called the "virtual route," was developed in TriTAPT; it permits analysts to see, for example, headway analysis and passenger load profiles on trunks shared by multiple routes and patterns.

The development of AVL-APC data collection and analysis capability poses numerous organization challenges. Perhaps the greatest is raising awareness within the organization of the value of archived AVL-APC data in order to ensure that AVL systems, which are often procured to serve real-time applications, have a design and the database support needed to achieve their potential for archived data analysis.

References

1. Furth, P. G. "Using Automatic Vehicle Monitoring Systems to Derive Transit Planning Data." In *Proceedings of the International Conference on Automatic Vehicle Location Systems*, Canadian Urban Transit Association, Ottawa, 1988, pp. 189–200.
2. Furth, P. G. *TCRP Synthesis of Transit Practice 34: Data Analysis for Bus Planning and Monitoring.* Transportation Research Board of the National Academies, Washington, D.C., 2000.
3. Cevallos, F. "Using dBASE to Collect AVL Data." *The dBase Developer's Bulletin 16*, July 2002.
4. Casey, R. F. *Advanced Public Transportation Systems Deployment in the United States: Update, January 1999.* Report FTA-MA-26-7007-99-1. FTA, U.S.DOT, 1999.
5. Casey, R. F., L. N. Labell, L. Moniz, J. W. Royal et al. *Advanced Public Transportation Systems: The State of the Art: Update 2000.* Report FTA-MA-26-7007-00-1. FTA, U.S.DOT, 2000.
6. Furth, P. G., T. H. J. Muller, J. G. Strathman, and B. Hemily. "Designing Automated Vehicle Location Systems for Archived Data Analysis." *Transportation Research Record: Journal of the Transportation Research Board, No. 1887,* Transportation Research Board of the National Academies, Washington, D.C., 2004, pp. 62–70.
7. Furth, P. G., J. G. Strathman, and B. Hemily. "Making Automatic Passenger Counts Mainstream: Accuracy, Balancing Algorithms, and Data Structures." *Transportation Research Record: Journal of the Transportation Research Board, No. 1927,* Transportation Research Board of the National Academies, Washington, D.C., 2005, pp. 207–216.
8. Furth, P. G. and T. H. J. Muller. "Service Reliability and Hidden Waiting Time: Insights from AVL Data." *Transportation Research Record: Journal of the Transportation Research Board,* Transportation Research Board of the National Academies, Washington, D.C., forthcoming.
9. Okunieff, P. E. *TCRP Synthesis of Transit Practice 24: AVL Systems for Bus Transit.* Transportation Research Board of the National Academies, Washington, D.C., 1997.
10. Furth, P. G., "Integration of Fareboxes with Other Electronic Devices on Transit Vehicles." *Transportation Research Record 1557,* TRB, National Research Council, Washington, D.C., 1996, pp. 21–27.
11. Kemp, J. Automatic Passenger Counting and Data Management System Project: Overview, Current Status, Lessons Learned. Presented at Orbital Users Group meeting, 2001.
12. Okunieff, P., T. Adams, and N. Neuerburg. *Best Practices for Using Geographic Data in Transit: A Location Referencing Guidebook.* Report FTA-NJ-26-7044-2003.1. FTA, U.S.DOT, 2005.
13. Lee, Y.-J., K. S. Chon, D. L. Hill, and N. Desai. "Effect of Automatic Vehicle Location on Schedule Adherence for Mass Transit Administration Bus System." *Transportation Research Record: Journal of the Transportation Research Board, No. 1760,* TRB, National Research Council, Washington, D.C., 2001, pp. 81–90.
14. Boyle, D. *TCRP Synthesis of Transit Practice 29: Passenger Counting Technologies and Procedures.* Transportation Research Board of the National Academies, Washington, D.C., 1998.
15. Barry, V. "Ongoing Use of Automatic Passenger Counters in Route Planning." In *Proceedings, Transit Planning Applications Conference,* 1993.
16. Furth, P. G., J. P. Attanucci, I. Burns, and N. H. Wilson. *Transit Data Collection Design Manual.* Report DOT-I-85-38. U.S.DOT, 1985.
17. Levy, D. and L. Lawrence. *The Use of Automatic Vehicle Location for Planning and Management Information.* STRP Report 4, Canadian Urban Transit Association, 1991.
18. Friedman, T. W. "The Evolution of Automatic Passenger Counters." In *Proceedings, Transit Planning Applications Conference,* 1993.
19. Navick, D. S. and P. G. Furth. "Estimating Passenger Miles, Origin-Destination Patterns, and Loads with Location-Stamped Farebox Data." *Transportation Research Record: Journal of the Transportation Research Board, No. 1799,* Transportation Research Board of the National Academies, Washington, D.C., 2002, pp. 107–113.
20. Barry, J. J., R. Newhouser, A. Rahbee, and S. Sayeda. "Origin and Destination Estimation in New York City with Automated Fare System Data." *Transportation Research Record: Journal of the Transportation Research Board, No. 1817,* Transportation Research Board of the National Academies, Washington, D.C., 2002, pp. 183–187.
21. Lehtonen, M. and R. Kulmala. "Benefits of Pilot Implementation of Public Transport Signal Priorities and Real-Time Passenger Information." *Transportation Research Record: Journal of the Transportation Research Board, No. 1799,* Transportation Research Board of the National Academies, Washington, D.C., 2002, pp. 18–25.
22. Schiavone, J. J. *TCRP Report 43: Understanding and Applying Advanced On-Board Bus Electronics,* Transportation Research Board of the National Academies, Washington, D.C., 1999.
23. Surface Vehicle Recommended Practice for Serial Data Communications between Microcomputer Systems in Heavy-Duty Vehicle Applications. Standard SAE 1708, Society of Automotive Engineers, 1993.
24. Surface Vehicle Recommended Practice for Electronic Data Interchange between Microcomputer Systems in Heavy-Duty Vehicle Applications. Standard SAE 1587, Society of Automotive Engineers, 1996.

25. VIGGEN Corporation. *FTA National Transit GIS: Data Standards, Guidelines, and Recommended Practices.* Report DTRS57-95-P-80861. U.S.DOT Volpe Center, 1996.
26. Kittleson & Associates, Inc.; Urbitran, Inc.; LKC Consulting Services, Inc.; MORPACE International, Inc. et al. *TCRP Report 88: A Guidebook for Developing a Transit Performance-Measurement System.* Transportation Research Board of the National Academies, Washington, D.C., 2003.
27. Kittleson & Associates, Inc.; KFH Group, Inc.; Parsons Brinckerhoff Quade & Douglass, Inc.; and Katherine Hunter-Zaworski. *TCRP Report 100: Transit Capacity and Quality of Service Manual,* 2nd ed. Transportation Research Board of the National Academies, Washington, D.C., 2003.
28. Strathman, J. G., T. J. Kimpel, K. J. Dueker, R. L. Gerhart and S. Callas. "Evaluation of Transit Operations: Data Applications of Tri-Met's Automated Bus Dispatching System." *Transportation,* 29, 2002, pp. 321–345.
29. Strathman, J. G., T. J. Kimpel, K. J. Dueker, R. L. Gerhart, K. Turner, D. Griffin and S. Callas. "Bus Transit Operations Control: Review and an Experiment Involving Tri-Met's Automated Bus Dispatching System." *Journal of Public Transportation,* 4, 2001, pp. 1–26.
30. Kimpel, T. J., J. G. Strathman, R. L. Bertini, and S. Callas. "Analysis of Transit Signal Priority Using Archived Tri-Met Bus Dispatch System Data." *Transportation Research Record: Journal of the Transportation Research Board, No. 1925,* Transportation Research Board of the National Academies, Washington, D.C., 2005, pp. 156–166.
31. Muller, T. H. J. and P. G. Furth. "Integrating Bus Service Planning with Analysis, Operational Control, and Performance Monitoring." In *Proc., Intelligent Transportation Society of America Annual Meeting,* Boston, 2000.
32. Furth, P. G. and T. H. J. Muller, "Conditional Bus Priority at Signalized Intersections: Better Service Quality with Less Traffic Disruption." *Transportation Research Record: Journal of the Transportation Research Board, No. 1731,* TRB, National Research Council, Washington, D.C., 2000, pp. 23–30.
33. Furth, P. G. and F. B. Day. "Transit Routing and Scheduling Strategies for Heavy-Demand Corridors." *Transportation Research Record 1011,* TRB, National Research Council, Washington, D.C., 1985, pp. 23–26.
34. Ceder, A. "Bus Timetables with Even Passenger Loads as Opposed to Even Headways." *Transportation Research Record: Journal of the Transportation Research Board, No. 1760,* TRB, National Research Council, Washington, D.C., 2001, p. 3–9.
35. Wilson, N. H. M., D. Nelson, A. Palmere, T. H. Grayson, and C. Cederquist. "Service-Quality Monitoring for High-Frequency Transit Lines." *Transportation Research Record 1349,* TRB, National Research Council, Washington, D.C., 1992, pp. 3–11.
36. AFNOR Certification. Service de Transport Urbain de Voyageurs: Reglement de Certification. Standard NF 286, 2002. http://www.afnor.fr, accessed May 1, 2005.
37. Kimpel, T. J., J. G. Strathman, D. Griffin, S. Callas, and R. L. Gerhart. "Automatic Passenger Counter Evaluation: Implications for National Transit Database Reporting." *Transportation Research Record: Journal of the Transportation Research Board, No. 1835,* Transportation Research Board of the National Academies, Washington, D.C., 2003, pp. 93–100.
38. Kemp, J. Archived Data Services: Lessons Learned–Things You Didn't Have to Think About When It Was Just AVL. Presented at 81st Annual Meeting of the Transportation Research Board, Washington, D.C., 2002.
39. Sampling Procedures for Obtaining Fixed Route Bus Operation Data Required Under the Section 15 Reporting System. Circular 2710.1A. FTA, U.S.DOT, Washington, D.C., 1990.
40. Townes, M. Letter to Acting FTA Administrator H. Walker. National Academy of Sciences Committee for the National Transit Database Study. June 1, 2001.
41. Evaluation of the Advanced Operating System of the Ann Arbor Transit Authority: Transfer and On-Time Performance Study: Before and After AOS Implementation. Electronic report EDL# 13147, ITS Office, U.S.DOT.

APPENDIXES

The following appendixes are published as part of *TCRP Web Document 23* (available on the TRB web site: www.trb.org):

- Appendix A: Tri-Met's Experience with Automatic Passenger Counter (APC) and Automatic Vehicle Location (AVL) Systems
- Appendix B: New Jersey Transit's Developing APC and Archived Data User Service
- Appendix C: AVL- and APC-Related Data Systems at King County Metro
- Appendix D: The Chicago Transit Authority's Experience with Acquiring and Analyzing Automated Passenger and Operations Data
- Appendix E: The Société de Transport de Montréal's Experience with APC Data
- Appendix F: OC Transpo: A Pioneer in APC Use for Service Improvement
- Appendix G: Hermes-Eindhoven's Experience with Automatic Data Collection, Operations Control, Management Information, and Passenger Information Systems
- Appendix H: The Transit Management Information System of HTM, The Hague
- Appendix I: Metro Transit's Integrated AVL-APC System